CULTIVATING ENVIRONMENTAL JUSTICE

CULTIVATING ENVIRONMENTAL JUSTICE
A Literary History of U.S. Garden Writing

ROBERT S. EMMETT

UNIVERSITY OF MASSACHUSETTS PRESS
Amherst and Boston

Copyright © 2016 by University of Massachusetts Press
All rights reserved
Printed in the United States of America

ISBN 978-1-62534-204-1 (hardcover), 205-8 (paperback)

Designed by Sally Nichols
Set in Adobe Minion Pro
Printed and bound by Sheridan Books, Inc.

Library of Congress Cataloging-in-Publication Data

Names: Emmett, Robert S., 1979– author.
Title: Cultivating environmental justice : a literary history of U.S. garden writing / Robert S. Emmett.
Description: Amherst : University of Massachusetts Press, 2016. | Includes bibliographical references and index.
Identifiers: LCCN 2016004214| ISBN 9781625342058 (pbk. : alk. paper) | ISBN 9781625342041 (hardcover : alk. paper)
Subjects: LCSH: Horticultural literature—United States—History.
Classification: LCC SB318.34.U6 E46 2016 | DDC 635.0973—dc23
LC record available at http://lccn.loc.gov/2016004214

British Library Cataloguing-in-Publication Data
A catalogue record for this book is available from the British Library.

This book is dedicated to Jack and Elisabeth Rose, with much love and hope that they will work for more just environmental futures.

CONTENTS

Acknowledgments ix

Introduction 1

1. The Democratic Roots of Twentieth-Century U.S. Garden Writing 11
2. Postwar Garden Writing, Literary Cultivation, and Environmentalism 44
3. *Being There, Second Nature,* and the Gardener as Pragmatist 67
4. Race, Regionalism, and the Emergence of Environmental Justice in Southern Gardens 102
5. Postindustrial America and the Rise of Community Gardens 132
6. Seeding New Territories 168

Epilogue. Garden Writing and a Phenology of Survival 185

Notes 199
Works Cited 207
Index 220

ACKNOWLEDGMENTS

This book resulted from a collective effort that began at the University of Wisconsin and benefited from the generous criticism and mentoring of Bill Cronon, Rob Nixon, Nancy Langston, Tom Schaub, Brad Hughes, Anne McClintock, and Lynn Keller. The community at the Nelson Institute Center for Culture, History, and Environment in Madison is responsible for the set of curiosities about cultural creativity, landscape transformation, and social justice that frame this project. Gregg Mitman's wide-ranging work on health disparities, environmental film, and the history of ecology has been a particular inspiration. Much love goes to Madison and CHE friends Peter Boger, Michelle Niemann, Andrew Stuhl, Heather Swan, and Anna Zeide and to the original members of the Eastside Writer's Group: Michelle Sizemore, Michelle Gordon, Travis Foster, Samaa Abdurraqib, and Lauren Vedal.

Colleagues in environmental studies at Oshkosh helped incubate ideas and opened new avenues for looking at urban and community gardens: Jim Feldman, Douglas Haynes, Stephanie Spehar, Paul van Auken, and students in my Environmental Justice seminar, the first but not the last class I took to Milwaukee to shovel compost and goat manure at Growing Power, Inc.

My cogardeners in Wisconsin also deserve thanks and seeds: Andy Dolan, Alissa Karnaky, Kristin Riching, Mike Darnell, Kate Nelson, and Mai Phillips. Gretchen Mead of the Victory Garden Initiative further agitated to get me out of my own backyard.

From my time in Milwaukee, thanks go to Kathleen Casey, Laura Murphy, Alan Singer, Robin Wiegert, and my students in the honors college who helped me explore the city and elsewhere as part of my U.S. Cities seminar.

In finalizing the manuscript I had the incredible good fortune to work at the Rachel Carson Center for Environment and Society in Munich, Germany. Starting with Christof Mauch and Helmuth Trischler, the list of the center's staff and international fellows who contributed to broadening the perspective of this book's themes grew by the month. I am especially grateful to my colleagues Katie Ritson, Arielle Helmick, Anna Rühl, Ursula Münster, Kim Coulter, Antonia Mehnert, Agnes Kneitz, Marcus Andreas, Carmen Dines, and Marie Heinz. Carson fellows Hal Crimmel, Thom van Dooren, Shen Hou, Tom Lekan, Celia Lowe, Jon Mathieu, Kenichi Matsui, John Meyer, Cameron Muir, Emily O'Gorman, Cindy Ott, Jenny Price, Laura Sayre, Nicole Seymour, Sai Suryanarayanan, and many more modeled intellectual rigor and commitment. Graduate students in the Environment and Society program (especially Elena Ruiz, writing on Detroit's urban agriculture) helped me grasp how aspects of American environmental and cultural history appear to those beyond U.S. borders.

European colleagues also rattled the bars of my training as an Americanist in useful ways. Thanks go to Marco Armiero, Hannes Bergthaller, Axel Goodbody, Tobijn de Grauww, Steven Hartman, Reinhard Henning, Dolly Jörgensen, Kati Lindström, David Nye, Ulrike Plath, and many others in Munich, Stockholm, and elsewhere.

Brian Halley has been the most professional and attentive of editors, for which I am grateful. The editing staff (including Mary Ellen Wilson) at the University of Massachusetts Press and anonymous reviewers gave generous feedback and curbed the worst of my excesses.

My parents and sisters provided moral and material support and served as examples of persistence and good humor. In some ways I owe even greater debts to my family-in-law, working academics who read and discussed drafts and ideas in this book: Ann LaBerge, Leigh Claire LaBerge, and Bernie Dukore.

My deepest gratitude goes to my wife, Louisa LaBerge, for accompanying me through the long process of researching and writing this book, reading every word, and encouraging many late nights of revision.

Portions of chapter 5 originally appeared in the journal *Interdisciplinary Studies in Literature and Environment* and are reprinted by permission of Oxford University Press.

CULTIVATING ENVIRONMENTAL JUSTICE

INTRODUCTION

In April 1990, *Harper's* magazine published a forum, "Only Man's Presence Can Save Nature," on the state of American public debate over the environment. In his opening comments, the moderator, Michael Pollan, proposed a thought experiment about an imaginary New England town's forest, the Tabernacle Pines: "This forest is extraordinary, with trees more than 150 feet tall. A hurricane came through recently and devastated the forest. Seventy percent of the trees are down. The place is a mess, almost impassable. I am curator of this forest, and I have to make a recommendation to the town. Do we leave it as it is—is that the state of nature?—or do we clear it out and replant pine, so that the next generation might enjoy some semblance of the old forest?" (38).

In response, Dave Foreman (of the advocacy group Earth First!) recommended that the town leave the trees to provide "even more services for the next 500 years—to beetles, termites, and fungi." James Lovelock, best known for his Gaia hypothesis, suggested that replanting may be necessary: "If the Tabernacle Pines were an *island* of trees . . . a kind of garden . . . [then it] might be more proper—even natural—to replant and rebuild." Frederick Turner argued that, although "nature is not a manicured garden," it should be viewed as a "set of complicated feedback systems, constantly exchanging information," and an "open-ended system" in which "the most crucial element is the human species"; the only way of "violating nature" would be *not* to involve townspeople in the question of what to do with the windblown forest. The evolutionary biologist Daniel Botkin took the position that the townspeople, as New Englanders, would probably prefer a forest, so they ought to "let the area grow back through the particular stages of succession that the Pilgrims saw . . . gardening the forest and weeding out the exotics" (38–39). Each proposed a different plan, but all spoke of "gardening" nature as a necessary cultural response to environmental catastrophe. Gardening leads to managing land intensively by balancing competing stories and the political interests they represent. To the forum respondents, gardening represents a more effective paradigm for debates over the environment than wilderness.

Mining the political significance of the garden as metaphor is nothing new. Gardens have echoed through American discourse and literature as symbols of a perfected society at home in nature, reconciling with wildness and civilization, industry and innocence. Yet the *Harper's* forum proposes the garden not as a metaphor but as a procedural model and social activity, with a goal to establish not a prelapsarian utopia or pristine nature but, rather, positive environmental justice, or a better balance of amenities and risks between people and the environment. In deciding how to proceed in the face of crisis (a destroyed grove of trees), these imaginary "gardeners" consult the embeddedness of their decisions in historical landscapes and question the efficacy of their inherited language. Through writing the garden, in other words, they seek to cultivate environmental justice.

Yet the group of respondents has a remarkably limited demographic profile for staking claims to a universal value. The language of "man's presence" in the forum's title reminds us that the participants were in fact privileged, white, and male. What kind of nature would they save, and for whom? Would a garden-centered discourse of environmentalism really move the political debate at century's end out of the impasse of the jobs vs. environment polarization of the 1980s? Or would it more likely lead to preserving stands of old-growth forest at the heart of wealthy Connecticut suburbs? After all, gardens in the United States had long been promoted as a way to produce good citizens—in Cecilia Gowdy-Wygant's analysis, to "make immigrants less of a threat" and "mold the character of the working poor" (185). Yet the tension between democratic, universal aspiration in progressive ideals and actual, historical practices has not prevented gardening from repeatedly taking center stage in U.S. environmental thought. Therefore, a sustained consideration of the literature of gardening—not as an isolated analysis of genre but within the political and environmental contexts in which gardens and garden literature were created—is long overdue.

In *Cultivating Environmental Justice: A Literary History of U.S. Garden Writing*, I explore how implicit claims and, by the end of the twentieth century, a language of environmental justice emerged in the practices and discourses of gardening across a variety of cultivated landscapes in America from the 1930s to the 2000s. This book takes Pollan's challenge seriously: given the global environmental crisis—a slow-motion disaster that no one can in good conscience ignore—one of our premier challenges as critics is to elaborate a gardener's model of environmental praxis. Not only would such a practice resemble Pollan's garden-centered ethic of more caring use of nonhuman nature (which I investigate in chapter 3), but it must also meet broader requirements of social justice beyond imaginary (and relatively homogenous) New England small towns. Efforts to conceive an environmental ethics within both ecocriticism and environmental philosophy first gravitated toward uninhabited places, meditating on wilderness and the insights it has inspired. Consequently, writing that emerges from and engages with everyday gardens has long been overlooked. In preferring sublime nature, we become deaf to quieter stories about people changing land and lives, narrating and giving form to the daily work of cultivating their existence. Such work is always social and inherently environmental.

Part of my goal is to clarify relations between individual writers and the representational limits implied by inherited definitions of genre that appear via discursive and historical analysis. A narrow view of garden writing, which is sometimes evident in existing anthologies, overemphasizes the cultural significance of gardens as sites of affluence and leisure. Gardens in U.S. cities since the turn of the twentieth century, in particular, have also come to stand for an ambiguous mode of interracial economic charity that can erupt into more direct demands for social justice. By analyzing what is generic and particular in these garden stories, I seek to connect deeper historical trends and reopen questions at the heart of environmental justice: Who really belongs in America, and where do they fit in? Who is deserving of sympathy or aid? What does pursuing happiness really mean in a culture that has designed individualist economic competition into its cities and fields? Gardens provide surprisingly good listening stations for hearing plural answers to such questions.

Women have undertaken (and continue to do so, in the case of food gardening) the vast majority of the unpaid work of cultivating environmental justice in the United States and around the world. Yet only within the last two generations have scholars (beginning with Annette Kolodny, Ann Leighton [Isadore Smith], and Vera Norwood) recognized those traditions of writing about the environment that have been practiced most often by women or given serious attention to everyday spaces categorized as feminine. Gardeners and garden writers such as Elizabeth Lawrence, Katharine White, Alice Walker, and Helen and Scott Nearing, as well as dozens of community-garden leaders, a majority of whom are women, have engaged gardening as an aesthetic and ethical practice that joins social and environmental concerns. These writers cultivated and critiqued the political meaning of gardening in a variety of publications; their work was unavoidably political because it required legal and cultural institutions to materialize ideals and weigh the demands of diverse neighbors. By uncovering layers of buried solidarities pressed into service in the garden, I will show how part of the unacknowledged history of the environmental justice movement derives from social attachments articulated in terms of locale, region, and work as well as from the more familiar categories of race, class, and gender.

Garden writing engages dialectically with writing and the environment; it is shaped by, and in turn has shaped, literary conventions and cultural landscapes. When we acknowledge this relationship, then, tastes in gardening become visible alongside the divisions of labor that shaped the land, even if only through fragmented sources. For example, African American gardeners enslaved at Thomas Jefferson's Monticello regularly sold produce to the president's family. While Jefferson enthusiastically recorded experiments in new tree and plant introductions in his garden and farm books, the sale of "garden variety" produce consumed at the plantation appears only in the household accounts kept by his granddaughter Anne Cary Randolph. Between 1805 and 1808, the teenaged girl recorded the first names of prolific produce sellers at Monticello—Bagwell, Squire, and Ned—all men and heads of families. Yet it seems likely that women and children also had grown the cucumbers, spinach, and cabbage that these three men sold to feed their owner's white family.[1]

A later group of ecocritics has been driven by what Lawrence Buell called "ecojustice revisionism," or a strong interest in understanding environmental justice that revises the principles for considering which texts are relevant (*Future of Environmental Criticism* 98). Environmentalist concerns for species extinction, wilderness, climate change, and pollution of air, water, and soil converge with social justice concerns of fair housing, healthy workplaces, the policing of gendered and racialized spaces, climate risks, and equitable access to transportation. I seek to further this more encompassing notion of environment and justice by following the routes of gardening through urban, rural, and suburban sites, from city block and backyard to cultural region and imagined national landscape. I compare not just differences in local, regional, and national culture but also the ways that garden narratives operate across these geographic scales.

In this pragmatic critical approach that attends to cultivation wherever it appears, I have taken the path for ecocriticism marked out by Dana Phillips in his 2003 book *The Truth of Ecology*. In the practical wisdom and humor of gardeners we find an antidote to the romantic wish fulfillment of earlier ecocriticism; to adapt a phrase from Phillips on the discipline's pragmatic turn, these garden writers are "more imaginatively engaged with the earth as it is, urban wastelands, wildernesses, and all" (241). And because the interstices between territories are arguably where the most significant environmental decisions are made—a majority of Americans now live and work in exurban areas—I pay particular attention to the connections between and the boundaries of deindustrialized cities, sprawling suburbs, and postagrarian countrysides.

Just as Americans have tended patches of land from inner city to wilderness edge, writing about gardens that crosses city, country, and suburb has emerged from diverse social groups and practices of gardening and writing. I trace this pluralism in both nonfiction and fiction through a variety of landscapes and traditions, including subsistence homesteads, vernacular rural gardens, and urban community gardens. The writers I explore have engaged gardening practices through pastoral, autobiography, polemic, and ethnography.

Pastoral conventions are as common as they are complex. Gardening has been associated with a desire for a simpler life imagined in the past tense, hostility toward modern society and its new technologies and metropolises, and a desire to preserve and pass along specific garden forms ("old-timey gardens") and even plants (heirlooms and rare native species). Since William Empson's *Some Versions of the Pastoral*, published in 1950, an established critical literature has sought to track pastoral as a mode of aesthetic and political representation within national and regional literatures (Leo Marx, Elizabeth Harrison) as well as across literary genres. By subordinating literary form to landscape practice, I try to recapture the more volatile forces at work in gardens, forces that are sometimes articulated in pastoral language. For example, when Helen and Scott Nearing evoked their flight from New York City to a New England farm in *Living the Good Life* (1954), they foregrounded a political context for their pastoral vision: anticommunist blacklisting pushed them out of the city's publishing and lecture markets as much as they were pulled by the promise of a life closer to nature. I consider both fictional and nonfictional engagements with gardening that have contributed to cultivating environmental justice in the United States, primarily in the twentieth century: novels and manuals, autobiographical and reflective essays, legal cases and institutional records, ethnographic studies and photo journals, newspaper columns and activist blogs, polemics and political satires.

I seek to show how writing about gardens reproduces values and practices that shape environments, especially social equity and justice, and that are then transmitted across generations. Thus the temporal structure of each chapter is intergenerational, rather than a traditional periodization of literary movements. Garden writing favors continuity and memory over aesthetic experiment or stylistic rupture because, as Bernadette Cozart, a community-garden leader in Harlem, has noted, "it takes only two generations to break the cycle of knowledge" necessary to garden effectively (qtd. in Hynes 11). In the case of the gardeners whom Cozart trains, the grandparents' residual knowledge of gardening in the American South has not always been transmitted to their city-dwelling children and grandchildren.

Still, the logic of intergenerational exchange does not preclude change—if anything, gardens are sites where values are annually re-created to preserve land and people. Most gardeners must remake their plots with each season so that, to a degree, even the most conventional gardens are susceptible to reinvention. As if to prove this point, the garden writer Elizabeth Lawrence—an apparent traditionalist—often described her home garden as an "experiment" (*A Southern Garden* 118). Scientific training in horticulture and her own family traditions led Lawrence to keep scrupulous phenological records (of bloom times) and the genealogies of plants in her garden as a technique for deciding what to plant the next year, what to preserve, and what to remove. Garden knowledge can be forgotten in a few generations; but in that same time, the significance of gardening in America has been radically transformed rather than lost. By the start of the twenty-first century, urban and suburban gardeners increasingly connected their work to the values of environmental justice.

The narrative arc of *Cultivating Environmental Justice* moves from the scale of individual private garden practices to regional gardening to a broad national movement of community gardening. Chapter 1 grounds the book in the deeper history of how gardening, ideas of nature, and democratic culture developed in the United States. I then explore subsistence gardens in the 1930s, a leap forward in time and outward in scale, along the way aggregating garden sites and practices to confront the increasingly public meaning of American gardens in culture and politics. Geographically, I move from gardens in an increasingly postagrarian countryside in New England to rural and urban gardens across the South to networks of community gardens in postindustrial cities, concentrating in New York and Los Angeles, to regional and national projects. Following this path, we will visit victory gardens, the subsistence plots of Old Left socialists, roadside wildflowers that inspired one of the first public criticisms of DDT, interracial encounters at farmers-market stands in the Jim Crow South, and urban community gardens that became sites of guerilla theater in protests against neoliberal privatization.

In chapter 2, I compare the Nearings' *Living the Good Life* and Katharine S. White's *Onward and Upward in the Garden* (1979) within the context of earlier so-called how-to literature, the historical concept of gardening for subsistence, and the politics of categorizing a literary genre of garden writing. Both the Nearings and White left New York City for rural Vermont and Maine, yet they tended vastly different gardens for different ends. The Nearings migrated permanently to garden in order to eat, calling their gardening Tolstoyan "bread labor," whereas White maintained a town-and-country habit, wrote primarily about flower gardening and reviews of garden writing, and relied on a hired gardener, Henry Allen, to implement her designs.

Beyond this fundamental contrast in the division of labor, I distinguish between the hostile rejection of the city by the Nearings and White's ironic, more nuanced preference for country/city interchange to highlight two strands of environmentalist discourse that emerge in their garden writing. Both are better understood against the background of midcentury debates over subsistence farming and the meaning of small-scale agriculture to national culture and political economy. Where victory garden manuals and propaganda present a nationalist appropriation of gardening, the Nearings' *Living the Good Life* articulates a radical social meaning for gardening that anticipates many of the major concerns of modern environmentalism (concern with air, water, soil pollution; lament for loss of spiritual value with loss of nature; desire to conserve greenspace and distribute environmental amenities equitably). White's essays in part retreat into ornamental gardening and announce a new tradition of environmental writing that takes toxicity as its theme, what Lawrence Buell has called "toxic discourse" (*Writing for an Endangered World* 31). The essays have broader horizons than her own private garden. In her *New Yorker* columns, White considers the transmission of practices through seed catalogs, flower arranging contests, prescriptive manuals, and botanical guides. She sketched the outlines of a genre of American garden writing that contained elements of literary and cultural criticism and contributed to popular awareness about the toxicity of common herbicides.

Chapter 3 unravels the lessons of *Second Nature: A Gardener's Education* (1991), by Michael Pollan, in part by comparing the character of the gardeners in *Second Nature* with that of the protagonist in *Being There*, the novel and screenplay by Jerzy Kosinski (1970; 1979). Kosinski's critique of garden language as a political ideology prepares us to be skeptical readers of Pollan's cultivation narrative. I consider how the latter's cultivation of character contributes to a garden ethic and articulates possibilities for a wider politics of environmental justice. Like the Nearings' homesteading account, Pollan's garden story begins in a recently abandoned farm landscape. Where the Nearings sought to extricate themselves from what they saw as the worst in American politics—the subordination of society to militarist corporate profiteering—and White pursued environmentalist change through literary means, Pollan presents an uneasy reconciliation between these two viewpoints. Ultimately, Pollan suggests that gardening can be both model and metaphor for addressing environmental problems in an ethnically diverse democratic society. He characterizes the archetypal "gardener" in ways analogous to Richard Rorty's "liberal ironist" and offers the image of his garden as a model for multicultural pluralism, joining diverse gardens and gardeners and thus escaping the dead metaphors and their political consequences exposed by Kosinski.

Environmental justice, or the equitable distribution of risks and amenities among communities, may be easier to imagine through more collective forms of writing than the reflective garden essay. How do we care for land that, at its most productive and beautiful, is also tainted with legacies of racial violence? I develop three consecutive and chronological readings of this type of garden writing that address such questions of equity: chapter 4 addresses the racialization of the division of garden labor in Southern garden writing; chapters 5 and 6 examine environmental justice as a discourse emerging in urban community gardens across American cities in the late twentieth and twenty-first centuries.

Southern garden writing is an important regionalist response to the rapid industrialization of the twentieth century whose devastating environmental and social effects continue to harm local populations (as in the case of mountain-top removal mining). Southern garden manuals are regionalist on both the climatic and cultural level, reproducing reform-minded environmental conservation agendas and persistently racialized divisions of labor. In particular, such manuals register the anxieties and possibilities of a changing social order in southern society as garden design wrestles with legacies of plantation landscapes. In Julia Lester Dillon's garden manual from 1922, for example, she describes how a boxwood hedge in a former plantation is worked into the design of the new owner's garden (43). A photograph of the hedge, plantation house, and two young girls becomes an iconic setting for southern gardens in black and white, exposing some of the ways in which social power has been ordered in landscapes through racial categories. If Dillon's contribution is "spreading the hedge," or perpetuating a symbol and tool of privacy, property, and social difference, later writers on southern gardens, such as Elizabeth Lawrence and Alice Walker, emphasize the leveling effects of plant exchange, breaking down boundaries between gardeners and potentially overcoming a regional project of racial signification as well as landscape design. I look closely at Lawrence's half-century-long career in the longer context of critiques of southern gardens by writers ranging from Frederick Douglass through Alice Walker. Lawrence's heuristic phrase "gardening for love," which she uses to describe how gardeners across the South, mainly farm women, exchanged advice, plants, and life histories in market bulletins, is haunted by her troubled awareness of the uneven exchanges across the permeable color line.

The fifth chapter explores the problematic of environmental justice by comparing a diverse archive of community garden stories. I look at organizational records, legal cases, ethnographies of gardens appropriated by urban gardeners, fictional depictions of community gardens, and polemics on behalf of gardens on public land bulldozed during the speculative real estate boom of the 1990s. These stories expose the contradictions in the use and ownership of urban public lands, renewing a mode of political writing that fuses claims to justice and demands to remake city neighborhoods—writing that Michael Denning, in his 1996 book *The Cultural Front,* called the "ghetto pastoral." I recount the contested legal history and polemical eyewitness accounts of struggles over community gardens on public land and analyze the ironic portrayal of community gardens in Jonathan Franzen's acclaimed 2001 novel *The Corrections*. The terms in which gardeners have articulated political identity and a right to cultivate contrast starkly with Franzen's critique of community gardens as the work of paternalistic affluent whites. Although exaggerated political claims about community gardens make them an easy mark for cynics, such spaces have always required difficult, collective, and sustained work by establishing the conditions and representing the material achievement of greater environmental justice. They represent a long-term, direct involvement of diverse publics in their making, interpretation, and continual remaking.

The final chapter of *Cultivating Environmental Justice* studies the convergence of postindustrial gardening narratives with more planetary perspectives and an appraisal of economic injustice. In the postindustrial garden narrative, cultivation as physical work and as a transformation of one's understanding of society involves restoring a healthy living environment while eschewing romantic notions of recapturing a pristine nature. Twenty-first-century garden writing has also migrated online and to social media; in the case of Novella Carpenter's *Farm City* (2009), it articulates a discourse of economic as well as ecological solidarity. This work of writing gardens is now about cultivating environmental justice by connecting us to abiding social bonds—or buried solidarities. In a time of economic and environmental inequalities, learning from more public-spirited and experimental writings about gardens seems urgent for literary critics and environmental historians, as well as for everyone concerned with the future of public spaces in the United States and beyond.

THE DEMOCRATIC ROOTS OF TWENTIETH-CENTURY U.S. GARDEN WRITING

Observers of gardens in the United States have long sought to square democratic hopes in a changing society that could neither guarantee all citizens access to land nor fulfill all the promises associated with intensively cultivating home grounds. From Jefferson's ideal of an agrarian republic of free land-owning farmers to Gilded Age relief gardens and Progressive Era school-garden programs, writers have linked gardens to their own broader social and political aspirations. Gardens became incubators and screens upon which they projected visions of democratic culture and allied ideas of nature. From the beginning of the nineteenth century to the end of the Progressive Era, gardens represented an ideal combination of natural aesthetics and democratic politics; that is, the selection and arrangement of plants described as natural were mapped on to virtues and institutions of sharing power between citizens in a republic. The historical peculiarity of this remarkably consistent ideal becomes apparent when we take a longer view of gardens in America.

For many centuries before the arrival of Europeans to the North American continent, indigenous cultures had codeveloped with horticultural practices, a relationship that continues today. Through successive generations the Seneca, Powhatan, Cherokee, Menominee, Mandan, Arikara, Hopi, and Pima, among many others, selected seeds of beans, corn, squash, sunflowers, and tobacco adapted not only to local soil and climate conditions but also to seasonal weather variations, resulting in varieties of crops with increased drought tolerance or disease resistance. Some tribes, such as the Ho-Chunk in the Midwest, designated clans as the tribe's seed keepers. The contemporary global diet owes a thousand-year seed debt to these gardeners, a debt attenuated because, except in rare instances (such as Gilbert Wilson's ethnographic work with the Hidatsa gardener Maxi'diwiac, called Buffalo Bird Woman), European Americans chose not to record the names of the individuals who transmitted the seeds and knowledge needed to make the settlers' gardens thrive. Decades of ethnographic and ethnobotanical research, notably the extensive work on Native American groups' seed saving, agroecology, and lifeways by the anthropologist Gary Paul Nabhan, have provided a window into the extent of horticultural practice and ecological knowledge during the millennia before European invasion.

Legacies of theft, appropriation, and transformation of the environment are never far from the idealized version of cultivation presented as a foundational national myth by European colonists. The political and environmental work of colonial cultivation is most apparent in documented cases of European immigrants assuming the gardens and territories of Native American tribes. Fields taken by force (as in the Finger Lakes region of New York) or vacated by disease (as in Tidewater Virginia and, later, along the Missouri River) furnished the first open land occupied by colonists. The earliest successful harvests of the Massachusetts Bay Colony were the progeny of stolen Indian corn planted in native villages emptied of their original inhabitants by newly introduced European diseases.

Beginning in the seventeenth century, French, Dutch, and English settlers grew subsistence gardens for "meate and medicine," and then the newly named "Americans" produced kitchen and dooryard gardens for "use and delight," culminating in grander landscape designs dependent on and reflective of "comfort and affluence" (Leighton 16). In her three-volume history of American gardens from colonial times to the end of the nineteenth century, Ann Leighton traces a commitment to "natural gardening" from the first seedsmen to the professional designers and scientists of the nascent Progressive Era. Although the focus here is on the century following the Progressive Era, the history of evolving and distinctively American garden ideas, by writers ranging from Bernard M'Mahon to Liberty Hyde Bailey, is worth revisiting with a fresh perspective, for it serves as background for what came later. These writers and their successors through the 1930s saw garden work as both imitating nature and cultivating democracy. Nature provided the template for a successful garden as well as an abstract ideal to be preserved and protected by the gardener's practice, eventually realized in the national landscape through the energetic participation of garden advocates and designers within the movement to designate national parks in the late nineteenth century.

Gardens and Democratic Culture from the Agrarian Republic to the Progressive Era

Gardens nourished millions of people in North America long before Europeans brought to the colonies their seventeenth-century almanacs along with their family bibles. These annual handbooks represent the earliest garden publications in English in the colonies and early republic as well. They contained excerpts from Virgil's *Georgics*, Gerard's Elizabethan herbal, and, later, borrowings from English advocates of scientific agriculture such as Arthur Young. The environmental historian Carolyn Merchant has argued that the imported almanac also furnished ecological knowledge for European colonists in North America (133). Merchant interprets the organic, holistic spiritualism of the almanacs as a set of preindustrial, nonmarket values that sustained a flourishing agroecological society of gardens, farms, and towns in the Mid-Atlantic and New England until the mid-nineteenth century. The incorporation of these regions into export markets, hastened by railroad and canal building and the crisis of the Civil War, ripped apart that older network of subsistence in what Merchant calls an "ecological revolution," a second revolution, in fact, after the systematic replacement of Native American horticulture and forest-based subsistence with the neo-European agricultural landscape.[1]

The astronomical calendars, pithy proverbs, and planting advice of the popular almanacs stabilized the weekday lives of farmers from Pennsylvania to Maine, but the first American garden books that drew the attention of political elites promised agricultural success and greater economic power for an emerging nation. Bernard M'Mahon, an Irish immigrant and professional nurseryman in Philadelphia, published the earliest American garden book, *The American Gardener's Calendar*, in 1806. John Randolph had published *A Treatise on Gardening* thirteen years earlier in Virginia, but his book offered only a modest collection of plant descriptions and a calendar modeled closely on British garden books. M'Mahon's more original *Calendar* went through eleven editions in the first half of the nineteenth century, making it the most successful and widely read American garden book of the period. His planting guide would eventually list over 3,700 available varieties and provided practical information about native cultivable plants as well as suggestions in a season-by-season guide for adapting European domestic crops to regional climate conditions. President Jefferson corresponded with M'Mahon after receiving a copy of the book.[2] In Jefferson's letter acknowledging the gift dated April 25, 1806, he requested to be sent "anything in that way hereafter curious or valuable" and regularly ordered seeds from M'Mahon's Philadelphia nursery, reflecting a shared enthusiasm for introducing new plant varieties.

From its beginnings, American garden writing succeeded through cultural and political prescriptions as much as for agroecological insights; literary and practical garden writing ratified colonial and then national narratives of development. Two decades before *The American Gardener's Calendar* was published, M'Mahon's contemporary Jean Hector St. John de Crèvecoeur had framed American agriculture within powerful physiocratic metaphors of environmental adaptation in his *Letters from an American Farmer*. Crèvecoeur's transplanted Europeans became Americans by flourishing in their new environment of benign colonial governance and access to good land, like so many young plants transplanted to rich new soil (69). In a similar vein, M'Mahon emphasized in his preface that "rapid progress in Gardening" ought to be "expected from an intelligent, happy and independent people, possessed so universally of landed property, unoppressed by taxation or tithes, and blessed with consequent comfort and affluence" (v). This first generation of European American garden writers framed gardening as at once thoroughly political (emphasizing liberal institutions, including popular literacy and freedom from interference) and environmental (transplanting robust stock and new varieties into rich soils in a benign climate).

In *Founding Gardeners* (2011), the British garden writer Andrea Wulf explores the centrality of gardening as an engine of economic development and wellspring of metaphors for the political ideals of four of the first five U.S. presidents. Wulf follows the private correspondence of Washington, Adams, Jefferson, and Madison as they describe garden cultivation and their home grounds, which they designed as didactic political landscapes.

Like M'Mahon, who inscribed his book "to the citizens of the United States," they saw gardening as central to the welfare of a new republic (10). Washington, Jefferson, and Madison planted native trees, tested improved agricultural methods and technologies, and laid out beds of ornamental vegetables at Mount Vernon, Monticello, and Montpelier in a naturalizing vision of an agrarian republic. Wulf celebrates the manifold connections of gardening and U.S. democratic ideals, but these connections also provide evidence for a garden ideology in the context of longer-term political and environmental transformations.

A more critical view of Jefferson's systematic record of his gardening experiments at Monticello reveals it to be an ambivalent archive of material bondage and half-enlightened republican ideals. The many tables of tasks and enslaved work crews in his garden and farm books reflect that Jefferson's enthusiastic embrace of horticulture as the best pursuit for an independent nation depended on the knowledge, skills, and appropriated labor of enslaved families. Monticello became a site for historical garden restoration when the Garden Club of Virginia replanted its flower gardens in 1939. Subsequent archeological excavations of former vegetable plots in 1979 led to a full-scale restoration that has continued under the project's present director, Peter Hatch. He published an account of the restoration work that emphasizes the significant presence of gardens and garden knowledge cultivated by black farm workers at Monticello (Hatch 64). Jefferson's gardens in Virginia served as seedbeds for transforming the nation's orchards, fields, and gardens, but Monticello also became a model for plantation landscape architecture in the South, with its gardenesque model slave quarters on Mulberry Row. The long invisibility of enslaved and free black garden laborers, particularly in the history of southern gardens, ought to temper how far we credit the claim of early nineteenth century seedsmen and presidents alike that gardens and democratic virtues were somehow linked in the early Republic.

In the colonies and early Republic, garden writing did not exist as a literary genre distinct from the political process of asserting land claims and botanical exploration. Puritan leaders, including John Winthrop, brought English garden manuals with them to the Massachusetts Bay Colony, manuals that instructed them to transform the landscape into a patchwork of fenced-off gardens and fields. As William Cronon showed in his classic environmental history of New England, *Changes in the Land*, they enacted this garden literacy into an improvement of nature that their courts would in turn interpret as the (divine) validation of land ownership (63, 130). The historian Patricia Seed subsequently demonstrated in a comparison with Spanish and Portuguese colonizers how the English planted gardens from Maine to the Caribbean as a "symbol of possession," where physically marking off and planting gardens validated English ownership (29). The availability of garden materials, new seeds and plants, moreover depended upon political projects of botanical exploration. President Jefferson entrusted M'Mahon and Benjamin Bartram in Philadelphia to preserve the collected discoveries from the Lewis and Clark expedition by cataloging, cultivating, and publishing the resulting information. It took another generation of botanical collectors to achieve that goal in tandem with militarized exploration of the West. Botany was more than a secondary project of military expansion, witnessed, for example, in the dozens of Californian plants named for the explorer John C. Frémont. Botanical collection abetted U.S. territorial expansion and built directly on the role garden literacy and garden making played in other British colonies. Garden writing, moreover, would make this Euro-American expansion appear both democratic and natural.

As early as 1822, the immigrant landscape gardener André Parmentier announced the superiority of a natural landscape style for gardens in the United States (Leighton 123). Parmentier's view of garden landscape design was elaborated by his successor in the Hudson River valley, Andrew Jackson Downing, who in turn borrowed from the leading British theorists John C. and Jane Loudon. Where gardens promised independence and improvement in the early republic, by the middle of the nineteenth century, a first generation of professional American garden designers was publishing refined ideas of gardens befitting an independent nation expanding in a lush environment. Downing, son of a nurseryman in Newburg, New York, became a self-taught garden designer who drew from a mix of sources in English; through study, designing his own estate, and travel, he absorbed the design and philosophy of British and German Romantic landscape design. The era's two leading designers, Downing and Frederick Law Olmsted, succeeded through persuasive writing and only then gained public commissions: Downing for the Capitol grounds and area around the new Smithsonian Institute, Olmsted for dozens of growing American cities from California to New York. Both argued that gardens furthered civic virtue by bringing Americans into contact with nature designed to enliven the mind through its beauty, rather than the economic benefits of scientific agriculture.

Downing set down the philosophical case for American gardens as areas of moral uplift, spaces conspicuously public for their virtues even when designed around private dwellings, from picturesque estates to humbler cottages for working people. He clarified this philosophy in *The Horticulturalist*, the first garden periodical in America, writing in July 1847: "We are inclined to claim also, for horticultural pursuits, a political and moral influence vastly more significant and important than the mere gratification of the senses" (*Rural Essays* 13). In particular, he saw gardening as a key activity for developing attachments "to one spot of earth" (16) and settling the potential destructiveness that he saw in his fellow citizens' restlessness.

Returning from Europe in the revolutionary year 1848, having visited the great public gardens of Paris, Frankfurt, and Munich, Downing staged a fictional dialogue with himself in twin guises: as the "Editor" of the *Horticulturalist* and an American "Traveller." The unnamed "Traveller" describes how the "last few months' residence in Europe, with revolutions, tumult, bloodshed on every side, people continually crying for liberty," frame his reappraisal of the American scene (*Rural Essays* 138). In particular, he writes that the French and Germans, in their public provision of gardens "open to all classes of people," are more "republican" in their "customs of *social* life, than Americans" (139). Downing's repatriated American worries openly over emerging class barriers in the United States. Through this persona, he laments how, "with large professions of equality, I find my countrymen more and more inclined to raise up barriers of class, wealth, and fashion." Such class division he finds "quite unworthy of us" and concludes that "we owe it to ourselves and our republican professions, to set about establishing a larger and more fraternal spirit in our social life" (142). Downing's prescription for true republican social life: large public gardens in growing American cities to "soften the feverish unrest of business which seems to have possession of most Americans, body and soul" (144), by drawing in the entire population to mingle and breathe their fresh air.

In August 1851, Downing repeated many of these arguments in a piece of propaganda for the proposed "New-York Park": "This broad ground of popular refinement *must* be taken in republican America, for it belongs of right more truly here, than elsewhere. It is republican in its very idea and tendency." Specifically, Downing claims that a public park "raises up the working man to the same level of enjoyment with the man of leisure and accomplishment," for "every laborer is a possible gentleman." He ends the essay with an exhortation: "Open wide, therefore, the doors of your libraries and picture galleries, all ye true republicans! . . . Plant spacious parks in your cities, and unloose their gates as wide as the gates of morning to the whole people" (152). In his sweeping vision for large public green spaces, Downing naturalized European liberal ideals of human freedom as transmitted through Romantic garden design. Democratic aesthetics and political ideals arrived with a generous spirit in his writing. Until his early death in 1852, he used the pages of the *Horticulturalist* to advocate for the conjoined moral and aesthetic virtues of gardens.

Downing's inheritor of the design of New York's Central Park was Frederick Law Olmsted, who exerted a more extensive influence on the national landscape even though his reputation waned over the twentieth century. Olmsted championed Downing's vision on an even grander scale, effectively seeing landscapes that became America's first national parks through the protective eyes of a garden designer.[3] The guiding naturalizing aesthetic of wilderness management came to Olmsted from Downing, Calvert Vaux, and Parmentier as well as from such earlier British garden writers as Batty Langley and William Robinson. The latter group had shaped the British estates and European parks that Olmsted studied during an extended European tour undertaken while serving as superintendent of the Central Park project. Olmsted implemented Downing's ideas of natural landscape design as a political project of improvement through hundreds of parks from Manhattan to Milwaukee, Boston to Berkeley. Eventually, the spacious oak groves, rambling paths, and green lawns of Olmsted city parks reflected a vision that united public health, environmental quality, and historical awareness.

Speaking in Boston in 1870, Olmsted placed the history of designing Central Park on a trajectory from urban design in eighteenth-century Europe forward several generations into what he imagined as a future

Manhattan of many millions. His vision of the benefits of open green space for city residents of all classes was expansive:

> Consider that the New York Park and the Brooklyn Park are the only places in those associated cities where, in this eighteen hundred and seventieth year after Christ, you will find a body of Christians coming together, and with an evident glee in the prospect of coming together, all classes largely represented, with a common purpose, not at all intellectual, competitive with none, disposing to jealousy and spiritual or intellectual pride toward none, each individual adding by his mere presence to the pleasure of all others, all helping to the greater happiness of each. You may thus often see vast numbers of persons brought closely together, poor and rich, young and old, Jew and Gentile. ("Public Parks" 18)

Here, Olmsted echoes the democratic enthusiasm and even the poetic cadence of his contemporary and fellow New Yorker Walt Whitman, who had also looked out on a teeming variety of city life with a generous (if somewhat paternal) gaze. In their critical history of the making of Central Park, the historians Roy Rosenzweig and Elizabeth Blackmar identified the gulf between Olmsted's democratic ideals and the actual making and use of the park, beginning with the eviction of black and immigrant dwellers from the area and struggles over conditions and pay for the workers who built the reservoir and contoured its celebrated landscape (77, 177). Rosenzweig and Blackmar conclude that "celebrations of the park's democratic character [in newspaper reviews in the 1860s] masked class-specific patterns of use" (212). Clearly, Olmsted's rhetoric of bringing together citizens in a public space and finding a common purpose also functioned ideologically, orchestrating real power inequalities, democratic aspirations, and moral interpretation of heavily cultivated but naturalized landscapes.

Olmsted's influence on landscape design and urban planning in America throughout the Progressive Era is hard to overstate. The environmental historian Shen Hou has argued that the articulation of Olmsted's vision of the "city natural," by the influential circle of his predecessors that published *Garden and Forest* from 1888 to 1897, represented the beginning of modern American environmental concern. Hou has demonstrated how sustained, synthetic, and pragmatic environmental ideas emerged in the pages of the journal published by Olmsted's friends and colleagues Charles Eliot, the landscape architect, and Charles Sprague Sargent, the Harvard botanist with whom he designed the Arnold Arboretum.

The great American horticulturalist Liberty Hyde Bailey, who developed scientific expertise and a cultural vision for horticulture in rural life, was one of the magazine's more prolific contributors. For Downing, Olmsted, and the generation of progressive urban planners and scientists they inspired (many later associated with the City Beautiful movement that peaked in the first decade of the twentieth century), American home gardens and public parks signified patriotic success and the possibility of social mobility. This progressive gardening ideal attempted to tame the emerging contradictions of a liberal democracy built on rising industrial wealth and a social division of labor into human-sized narratives and city designs for livable, more just environments (Hou 151), again synthesizing democratic culture and garden design. As a discourse, it had its own ideological limitations.

The success of a progressive garden ideal in the United States built on the republican zeal of Madison, Jefferson, and Washington, which had elevated ornamental farms, protected woodlots, and improved horticulture as models of democratic virtue. Yet these same advocates of gardening in the early Republic depended on the industry, elitism, and ideas of scientific progress of the American industrial aristocracy, which directly funded the advance of scientific botany and horticulture. Many of the iconic public landscapes associated with Olmsted (Central Park, the Stanford University campus, and Niagara Falls) owe as much to the intervention of financial elites as they do to democratic ideals. A few of the first super-rich Americans commissioned garden estates from Olmsted to rival their European counterparts at Versailles, Blenheim, and Muskau.[4]

American garden writing of the Late Victorian period and Progressive Era was a genre produced by and for this liberal class of educated elites.[5] Many had been inculcated, like Olmsted, with a sense of moral obligation to the "vast numbers" of poor who could be brought into the "pure air and under the light of heaven" of a gardened public space, as he described Central Park and Prospect Park in 1870 ("Public Parks" 18, 19). A century later, only a minority of garden writers echoed Olmsted's public-spirited political conviction or the revolutionary republican spirit of Downing. In the intervening decades, an explosion of commercial publications catered to different literary tastes and quieter, depoliticized views of gardening. Seasonal, climate-specific design manuals, popular magazines, and reflective garden essays proliferated as scientists, sentimentalists, and adventurers published hundreds of titles between 1890 and 1930, aided by the growth of the American Garden Club (founded in 1913), the launch of national periodicals of a progressive, scientific mold (*Country Life* in 1901), and more commercial aims (*House and Garden* in 1911). Virginia Tuttle Clayton has argued that popular magazines during the Progressive Era, including *Ladies Home Journal,* touted an ideology of the "wild garden" particularly among middle-class amateur gardeners (153). These gardeners aligned themselves with nature as a positive force—a way to distinguish themselves from other classes and distasteful aspects of modern life. Popular magazines amplified an American garden ideal and melded scientific horticulture, political power, and moral discourse.

The most established individual scientific publications and systematic garden institutions in America all date to the Progressive Era: Liberty Hyde Bailey's first edition of *Hortus* as well as the Arnold Arboretum, Brooklyn Botanical Garden, and the land-grant extension agencies beginning with Bailey's initiative at Cornell University. This writing is more notable for the stability of its dominant tropes and moralizing than for literary innovation. It is conservative even in its praise of progress and respectable even when it attempts humor. For example, the adventurer and Harvard plant collector Ernest H. Wilson's plant expeditions on four continents yielded such titles as *Aristocrats of the Garden* (1917), wherein he indexed tree species with an extended metaphor that naturalized social stratification, from lowly common shrubs to stately elms and oaks. The individual specimen of the collector, the plant rather than the landscape, came into greater focus for Wilson's botanizing imagination. In discussing ornamental flowering cherries, he defended fruitless beauty, "since the fruits they produce have no comestible value we can drink in their charms uninfluenced by the pernicious alloy of utilitarianism" (196). Considering Wilson's commercial books alongside the first American garden publications reveals a shift of emphasis from republican virtue (introducing a new strain of commercial apple or pear) to private aims of security, status, and entertainment (introducing new species for ornamental flowering borders), a testament to both Wilson's distance from subsistence gardening and increased social stratification across the nation. With the ecological revolution of industrialization had come an end of agrarian, egalitarian society. An industrial, class-divided society did not realize a home and garden for all, but for a while its garden writers pursued that goal.

The first relief gardens on public land, destined for unemployed workers, also date to the dawn of the Progressive Era. Like *Aristocrats of the Garden,* Detroit's potato patches for poor workers and similar programs in New York and Philadelphia in the 1890s were symptoms and treatments of social inequality. The relief gardens are harbingers of environmental justice concerns in twentieth-century garden writing and hearken back to agrarian, republican ideals. In his prolific garden writing and championing of rural life, Liberty Hyde Bailey, Ernest Wilson's friend and correspondent, provides a more positive bridge to later environmental justice concerns. Bailey's evocative and visionary book *The Holy Earth* (1915) repays a close reading for how it imagined a protoenvironmentalist vision of civic responsibility set within the political economic context of horticulture and agriculture.

Bailey wrote of farmers and other "land workers" in New England who were unable to make a living and, increasingly indebted, could not hold on to their small farms. He also tells of a need for what he called "spiritual contact with nature," which he describes in terms of environmental connectedness and expressive, psychological individuation. The result of successful connection with nature and subsequent individual expression was the "separate soul" (130), an authentic and environmentally integrated self. Bailey recognized John Muir, the writer and founding president of the Sierra Club, as his exemplar. But rather than emphasizing the value of wild lands uniquely, Bailey sets up an explicit foreground/background relationship between cultivated land and wild lands (forests, open fields, and sea). He then remarks: "I wish that we might know the forest intimately and sensitively as a part of our background. I think it would do much to keep us close to the verities and the essentials" (155). Among these essentials, Bailey placed good materials, food, and intelligent management of the land.

Bailey voiced the sanctification of material life in language that evokes Jeffersonian agrarian democracy: "Merely to make the earth productive and to keep it clean and to bear a reverent regard for its products, is the special prerogative of a good agriculture and a good citizenry founded thereon." Good agriculture "carries the final healing" for the "mad impersonal and limitless havoc" he saw in the wider world. In a vision of agrarian culture's stability that would be taken up by successors such as Wendell Berry, Bailey argues that "while the land worker will bear much of the burden on his back he will also redeem the earth" (119). Bailey's writing career and institutional advocacy for horticultural research in universities, nature education at all levels of public education, and public agencies to secure the livelihood and quality of life for rural communities extended through two world wars and repeated economic crises. In this sense, he was the first American garden writer to witness the full social and environmental devastations of the twentieth century.

Bailey's late and revised writings on the idea of progress in horticulture provide a reflexive lens into this crisis in the progressive concept of gardening. In 1911, Macmillan published a sixth edition of his essays on evolution, *The Survival of the Unlike*, first printed in 1896. These he had dedicated to his mentor, the Harvard botanist Asa Gray, Darwin's most powerful advocate in the American scientific world. In the essay "Recent Progress in American Horticulture," Bailey writes with great ambivalence about the rapid change in gardens and gardening. On the one hand, he continued to see gardens as engines of growth and progress, noting that the "horticultural settlement of our great west and of the cold north is one of the wonders of our time," and that a continued increase in plant varieties would make "our horticulture . . . the richest in the world." He adds, "We can scarcely conceive what riches the future will bring" (211).

On the other hand, Bailey saw limits to methods he dubbed "progressive gardening" and signs of concern in the failing economics of agriculture, particularly. He devoted several pages to the "most conspicuous recent advancement . . . the advent of sprays for destroying insects and fungi." Although he praised more use of sprays, he emphasized that "we are only keeping pace with the initial progress fostered by the origination of new varieties and the quickening commercial life of our time." He ends the section with an eerily prescient image of the chemical warfare that would come with the First World War: "But in your generation and mine, men must shoulder their squirt-guns as our ancestors shouldered their muskets, and see only the promise of time when they shall be beaten into pruning-hooks and plowshares and there shall come the peace of a silent warfare!" (215). He argued that this future "peace of silent warfare" would come with the application of effective biological controls and rotation practices, "control[ling] natural agencies that one will counteract another," producing a "comparative equilibrium" distinct from early ecological thinking about natural equilibrium.[6] Bailey's critical assessment of pesticide use and cautionary tone anticipated by half a century Rachel Carson's *Silent Spring*.

A remarkable consistency characterizes the terms with which American writers praised gardening, from the first guide published in 1806 to the decade of deepest political, economic, and environmental crisis after 1929. Writers had celebrated garden work as good for the soul, a tonic of redeeming physical labor, and good discipline for potentially unruly children (and, later, potentially rebellious workers). Late Victorian garden writers such as Celia Thaxter, Alice Morse Earle, and Louise Beebe Wilder opened up garden writing to greater sensual appreciation of fragrance, color, and memory. Yet they also couched sensuality within the moral appeal of gardens. The organizers of the first national Garden Week in 1924 claimed gardening taught children a healthy work ethic and elemental spirituality (Unger 106). Such moral benefits echoed prescriptions from Catharine Beecher and Harriet Beecher Stowe's domestic labor tract of 1869, which recommends the civilizing influence of garden work for children and daughters (296). Both the organizers of Garden Week and their predecessors articulated home gardening as vital to democratic culture: cultivating a pea patch or pruning a pear tree served as a universal solvent for class tensions, domestic strife, and medical complaints. Stowe, Beecher, and then Olmsted and later leaders of the Progressive Era in the Garden Club movement, such as Louisa Yeomans King, also interpreted gardeners' love of the green world as a sympathy that crossed socioeconomic lines.[7]

At the national level, Progressive Era gardeners like Louisa Yeomans King perhaps had trouble seeing beyond the garden gate to the grittier realities of early twentieth-century cities and struggling rural districts. Immigrants from southern and eastern Europe who settled on New York's Lower East Side were limited to memories of gardens or token displays of potted plants, which affluent observers interpreted as signs of self-respect and moral principle. Observers not only coded gardening in terms of class tastes but also viewed them through the racial biases of dominant U.S. intellectual culture. In 1890, the Danish American reformer Jacob Riis contrasted German immigrants' gardening in the tenements of New York with what he saw as the damaging political power of their Irish neighbors. Riis argued that "the German has an advantage over his Celtic neighbor in his strong love for flowers, which not all the tenements on the East Side have power to smother." Wherever a German immigrant moves, according to Riis, he "turns his saloon into a shrubbery as soon as his back-yard," where "it does the work of a dozen police clubs" (124). Social reformers and urban critics at the end of the twentieth century, as we will see in chapter 5, would return to gardens on the Lower East Side as forms of cultural work, moral value, and a potentially more just (although not more policed) social order. They would also unconsciously reintroduce categories of social difference in the process of imagining and cultivating protected enclosures of vegetables, flowers, herbs, and fruit trees.

Natural Gardening and the Prairie Roots of Ecological Restoration

Writers and gardeners in the United States designed landscapes to reflect economic and political ideals of fairness, justice, and mobility; they produced self-justifying ideals, "discovering" their political values in the environment, then calling these ideals "natural." Nowhere is this historical, political process of ideal formation and projection clearer than in the emergence of practices labeled since the 1960s as ecological restoration. To a large extent, ecological restoration as a concept and practice developed in the 1930s from ideas about natural gardening in Great Britain, Australia, and North America. These ideas negotiate difficult ethical territory, for even the early American and British theorists of natural gardening identified it as an aesthetic, political practice associated with national traditions—at times, their praise for the "northern European" or "Anglo-Saxon" practice of natural gardening (language from Alexander Humboldt's *Cosmos* that Andrew Jackson Downing introduced to his readers) paralleled racist and nativist ideologies. Elsewhere, American writers would label impressive natural landscapes "gardens" to signal these areas were worthy of protection—a confusion of politics, botany, and secularized nature worship.

In 1912, when John Muir wrote about Fountain Lake, Wisconsin, in his memoir, *The Story of My Boyhood and Youth,* his memory stretched across decades in which most of the prairie had been plowed under for farms. He remembered kettle lakes filled with water lilies, *Nymphaea odorata,* and in his mind's eye "no lily garden in civilization we had ever seen could compare with our lake garden" (97). As a young man writing *My First Summer in the Sierra,* Muir would also describe the mountain range's riparian flora and flowering meadows as "gardens" (172, 206). Given the sacralization of nature in his writing, Muir likely meant *garden* in its Edenic sense. More generally, in describing water lilies, moist meadows, and fields of alpine wildflowers as gardens, Muir expressed a protective sentiment toward the landscape. Such language imagines territories reserved from anthropogenic damage, including browsing from the sheep Euro-Americans like Muir introduced to the Sierra, which he called "wooly locusts" but earned wages to protect (208).

Muir expressed sympathy for the rapacious hunger of his sheep and awareness that they furnished him with the means to spend his first summer in the Sierra, ascending into the high country in search of greener pastures while sketching, writing, and botanizing along the way. Muir's garden vision of prairies and mountain valleys supported his later arguments to protect entire landscapes, scaling up his habit of guarding a few panther lilies along the north fork of the Merced from the herd (185). In the case of the Yosemite, his Sierra Club succeeded in translating this vision into legal protection. The vast prairies of Muir's childhood suffered a different fate.

Historians and ecologists refer informally to the tall- and short-grass prairies, which once covered over two hundred million acres from northern Texas to Wisconsin, west from New Mexico up to Montana, as America's last settled and first wholly devastated ecosystem. Yet paradoxically the great grasslands produced the redemptive practice of ecological restoration, and with it a commitment of doing justice not only to other humans but also to the larger land community. Midwestern gardeners, horticulturists, and botanists invented the concept and practice of ecological restoration in Illinois and Wisconsin around 1930. Restoration ideas and Aldo Leopold's call to treat all members of the land community more justly developed in dialogue with those devoted to the "native landscape," a category that had become influential years earlier among gardeners and landscape architects.

The first great advocate of replanting native prairie wildflowers was a gardener, teacher, and writer, Jens Jensen. Like Muir, Jensen had immigrated to the Midwest from Europe and fallen in love with what remained of the wilder landscapes at the edge of a rapidly expanding Chicago metropolis. For practical, aesthetic, and ethical reasons, Jensen held native plants and landscapes to be superior to exotic plant materials and formal geometric designs. In his meditative essays collected as *Siftings* he notes: "Art must come from within, and the only source from which the art of landscaping can come is our native landscape. It cannot be imported from foreign shores and be our own" (61). Viewed in its biographical and historical context, Jensen's idea of "our native landscape" appears complex and ambivalent. Jensen's practical gardening work and park designs inspired the development of ecological restoration in the 1930s and have been largely overlooked in histories of U.S. environmental thought, making his imported nativism worthy of close analysis here.

Jensen emigrated from Denmark in 1885 and settled in Chicago, where as a young man he worked his way up from a day laborer and gardener in the West Parks Commission; by 1905 he had become the chief landscape architect and eventually served on the city council's Parks Commission. In these roles, Jensen designed significant individual parks and, in 1920, a parks plan for the city of Chicago (Tishler xii). The plan called for community food gardens adjacent to large public parks that preserved sand dunes, remnant prairie, and forest areas in Cook County. Jensen imported the natural garden aesthetic from English models, as had Downing two generations before. His preference for native plants seems particularly close to the natural garden ideal of his contemporary, the British garden writer William Robinson.

Robinson favored native plants, particularly hardy English perennials, grouped in informal, organic clusters rather than the dominant Victorian trend of bedding out annual flowers in geometric, color-coded patterns for a mass effect. His book *The English Flower Garden,* first published in 1883, revolutionized gardening in the English-speaking world, calling for wild gardens of naturalized native species and mixed perennial borders. From his position as gardener in the Royal Botanic Gardens, Regent's Park, Robinson sounded a populist note, seeking his models not from grand private estates but rather from the "lovely cottage gardens in the country round London" (viii). Although it had a popular flavor, above all the natural, or wild-garden, aesthetic aligned human creative work with the historical accident of native vegetation.

Gardening with local vegetation supported a vision that blurred nature and culture, human and nonhuman agency. Robinson articulated this as a universal creed succinctly in the foreword to the eighth edition of *The English Flower Garden*. There he wrote that the aim of garden design was "to make the garden a reflex of the beauty of the great garden of the world itself, and to prove that the true way to happiest design is not to have any stereotyped style for all flower gardens, but that the best kind of garden should arise out of its site and conditions as happily as a primrose out of a cool bank" (viii). He insisted that "the gardener is the trustee of a world of fair living things, to be kept with care and knowledge in necessary subordination to the conditions of his work" (7). These natural or wild gardens likely also caught on because of changing conditions of horticultural work at the turn of the twentieth century on English estates. A wild garden required fewer laborers to maintain and less capital than for repeated plantings of flowering annuals grown in hothouses.

Practical as well as pecuniary concerns also motivated the first large-scale planting of native perennials in a major U.S. public park: Jensen's "American Garden" in Union Park in 1888. After trying unsuccessfully to plant formal gardens of transplanted foreign plants from propagating beds, Jensen decided instead to relocate native perennial wildflowers from the edges of nearby woods. The landscape architect Robert E. Grese has described how Jensen's natural gardening style spread from popular successes in public parks to private patronage in his study *Jens Jensen: Maker of Natural Parks and Gardens* (1992). Jensen gained private commissions between 1910 and 1930 from prominent midwesterners including the Ford family and Julius Rosenwald, founder of Sears and Roebuck (99–103). Connections to wealthy patrons, in turn, increased the regional influence of Jensen's "prairie style" of natural gardening. Between 1890 and his retirement after 1930, Jensen designed more than fifty landscapes along Chicago's wealthy North Shore.

Jensen's garden designs and organization of conservation groups in Illinois and Wisconsin reflect his fusion of democratic values with a valuing of natural landscapes in the region. In 1913, Jensen founded Friends of Our Native Landscape (FONL), a club that organized outings to landscapes of exceptional natural features in the greater Chicago region, such as the Indiana Dunes and Wisconsin Dells. FONL staged outdoor theater in the open "player's green" at Columbus Park in Chicago and in the council rings that Jensen included in many park designs. These stone benches encircling a fire pit are a signature feature of Jensen's work and were usually positioned in wooded enclosures with a view into a meadow opening or lake panorama. Most are designed to seat a dozen visitors and create the intimacy of an outdoor room. Grese describes Jensen's council rings as a key feature of garden design: "Viewed by Jensen as a symbol of democracy, the council ring was intended as a gathering place where all people would be equal.... Around the council ring, people gathered to participate in free and honest discussions, to read poetry or tell stories, to act out dramas, or simply to meditate, especially on humanity's relationship to nature" (176). Thus through design and programming, Jensen used public gardens to dramatize his vision of preserving native landscapes.

From his friend, the Chicago playwright Kenneth Sawyer Goodman, Jensen requested a masque, which FONL members performed at annual gatherings in parks he had designed. In Goodman's masque, sometimes called "At the Edge of the Woods," allegorical characters represent the narrative of development common during the Progressive Era. A spirit of the "Ageless Beauty of the Wild" appeals to a series of figures intended as phases of prairie history: an Indian who appears, only to be addressed as an "unmindful child" (line 28) by the spirit, then is bid farewell as he is "hunted hence" (39); a Pioneer who drives the spirit from its path; a Builder who declares "wilderness—/ Just one damned aching nothing" (92–93) and curses the spirit's lack of "dollar-getting wits" (126); and finally the Friend, who pledges "protection for thy home" to the spirit of the wilds (163). Repeated FONL performances of Goodman's drama implored visitors to preserve the native landscape. Like other conservation organizations, such as the Woodcraft Indians and the Boy Scouts of America, FONL called for protecting the wild by summoning images of a vanished race in touch with nature. Shepard Krech III has described this phenomenon as the myth of the "Ecological Indian," an essentialist view of native peoples as inherently conservationist, which Krech calls "ultimately dehumanizing" because it denies cultural variability and ignores the complex ecological changes wrought by native tribes (26). Goodman's short play ritualized the historical exile of indigenous peoples as part of a progressive trajectory ending with more recent European immigrants speaking as self-styled friends of native landscapes. Thus a European-American gardener, Jens Jensen, also became paradoxically the apostle of "native landscape" as a place of spiritual value without native human inhabitants.

In *Siftings,* Jensen defended the spiritual as opposed to the commercial value of the river bluffs, oak clearings, and grasslands. In the essay "Compositions," printed posthumously in 1956, Jensen wrote of the bluffs along the upper Mississippi: "With few exceptions this land is unfit for food production and was primarily preserved for the soul—it was so destined. Primitive beauty still reigns supreme. It is America in primeval state and it is still here in a great measure. But even these secret shrines are not safe from spoilation. Nothing is safe that might bring forth the Dollar, the God of our age" (142). The vision here is complex, with shades of primitivism, fatalism, and an anticonsumer critique. In claiming that landscapes are in a "primeval state" and "primarily preserved for the soul," Jensen interprets nature both ahistorically and teleologically. The uncultivated land is "destined," like "secret shrines," for the purpose of sophisticated aesthetic and moral appreciation.

Not all friends of the native landscape shared Jensen's ahistorical sense. After retiring from his private landscape design firm in Illinois, Jensen moved north and became active in the Wisconsin chapter of FONL that he had founded in 1920. In the 1930s, the group included the ecologist Aldo Leopold, the botanist G. William Longenecker, and their fellow professor at the University of Wisconsin, the landscape architect Franz Aust. Along with a young botanist named John Curtis, members of this circle developed methods for restoring grassland communities, the first systematic work of vegetation ecology in the United States (Curtis's *Vegetation of Wisconsin* of 1959), and engaged in an ethical discourse rooted in a dynamic, historical perception of land, made famous by Leopold's posthumous *A Sand County Almanac.*

In his 2003 book on ecological restoration, *Nature by Design,* the philosopher Eric Higgs has rightly objected to the mythic tale of origins linking restoration to Leopold's lone "genius" (83–84). "If we extend a line between gardening and restoration," Higgs writes, "somewhere along the line, the border separating the two is going to become a matter of convention and judgment" (91). Leopold penciled in a "site for prairie amid a scattering of food patches for game birds on a rough map of the UW-Madison Arboretum in the fall of 1933," according to two leading practitioners of ecological restoration, William Jordan and George Lubick (80). They describe this moment as a result of a collective Wisconsin project: Longenecker, a professor of horticulture, designed the plantings, the botanist Norman Fassett and his students planted the first trials, and then Theodore Sperry was hired as the prairie ecologist (78–79). Over the next decades, Civilian Conservation Corps workers, sympathetic home gardeners, and arboretum volunteers labored to maintain the prairie through plantings, annual burning, and removal of invasive trees. As elsewhere in North America and Australia, ecological restoration theory followed gardening practice, adding new ideas of landscape-scale cultivation to those inaugurated with colonial botanical gardens and public parks

The invention of ecological restoration in the 1930s in North America also issued from the manifold crisis of the Great Depression and, in Wisconsin, from awareness of the ecological devastation of the dust bowl (Higgs 78). Jordan and Lubick describe ecological restoration as emerging when natural gardening aesthetics combined with ecological science. A trajectory runs from the aesthetic-spiritual vision of Downing, Olmsted, and Jensen to the ecological-community values of restorationists (45). Yet the tangled garden genealogy of ecological restoration is still thicker, with common roots in communitarian political philosophy: in Chicago, Jensen met the plant ecologist Henry Chandler Cowles (Jordan and Lubick 48). Cowles wrote the first papers on plant succession in the Indiana Dunes and was the first plant ecologist to offer classes in that field through the geography department at the University of Chicago. The emerging ecological sciences implied and developed from within accompanying social and political contexts, as historian Gregg Mitman makes clear in his 1992 study of Cowles's work and the Chicago school (17). From Cowles, Jensen learned about native plant communities and ecological dynamics, what was then referred to as the "sociology" of plants (Grese 52). The influence of vegetation ecology on Jensen's garden design and activism is in turn reflected in his descriptions of plant communities as associations of "friends." The "Friends of the Native Landscape" groups thus blended ecological science and communitarian values. Jensen led FONL outings to the Indiana Dunes and Cowles served as the first vice president. As contemporaries in Chicago, both clearly shared a democratic, progressive vision of communities based on harmonious cooperation rather than ruthless competition between individuals.

Gardening served as a practical and symbolic means to perpetuate progressive values and a hopeful vision in a time of general gloom, espousing ideals of just and fair communities in touch with the land, even as these ideals reproduced ethnocentric blind spots. As with later economic crises, the Great Depression created opportunities for experimentation by curious gardeners, horticulturists, and botanists. Bankrupt and badly eroded farms could be bought or leased for relatively little. Failed depression-era farms enabled gardening at a scale to re-create a semblance of the prairie ecosystem that Muir experienced in 1848, when he arrived in the United States. In 1933, Longenecker, Curtis, and teams of graduate students and laborers working for the Civilian Conservation Corps replanted acres of big bluestem, buffalo grass, and prairie wildflowers on a former farm acquired by the university in Madison (Jordan 77). Workers also seeded the fields with local rocks that had once been carefully removed by farmers. Systematic study of remnant long-grass prairie vegetation communities and oak savannah elsewhere in the state, along with years of experimentation, produced the world's first conscious effort in ecological restoration. Renamed Curtis Prairie, the site also represents the longest sustained restoration.

In 1935, the same year that Aldo Leopold secured a lease on a worn-out farm in Sauk County where he began to plant thousands of pine seedlings to hold the sandy soil, Jens Jensen broke ground on a new home in Ellison Bay, Wisconsin. With patronage from Edsel Ford, he transformed "The Clearing" into a folk school based on traditions in his native Denmark and a demonstration landscape for his ideas. Over the next few years, Longenecker would bring his botany students to the Clearing, and Jensen would lecture in the landscape architecture program in Madison. The Wisconsin chapter of Friends of Our Native Landscape survived until 2000, by which time ecological restoration had become a worldwide (and contested) principle of land management adopted by national governments and conservation NGOs, critiqued for its unintended consequences wrought on indigenous peoples.[9] Certainly the uneasy mixture of progressive social hopes, landscape design from above, and Eurocentric aestheticism in Jensen's work calls for further attention. His ideas of natural gardening have been reproduced in hundreds of restoration projects across the United States.

Food Gardening in the Era of Liberal Industrialization

The literary historian Timothy Sweet has characterized the twentieth century as a process of increasing "environmental alienation" as citizens found themselves living and working far from the landscapes that inspired a naturalized, national ideology.[10] Sociologists, geographers, environmental historians, and scholars in food studies have described a food system that increasingly concentrates capital in the hands of fewer and larger operations, with gardens offering marginal resistance in community food security and food justice movements.[11] More and more of Bailey's land workers left the country and a smaller share of immigrants found farms, so that by 1920, roughly half the U.S. population lived in cities. Garden writing of the generation that reached its prime between 1930 and 1970 engaged with the increasing environmental alienation of these urbanizing and suburbanizing conditions. At the same time that Americans increasingly made a living distant from agriculture, literary values in America also were shifting.

The less seriously gardens mattered in agricultural terms for American readers, the more emphatically garden writers emphasized their moral value in terms of fragrance, social status, and memory. A small but significant body of garden writing focused on subsistence agriculture and domestic economy in the 1930s, only to shift, in the period immediately following the Second World War, to a consumerist focus on aesthetic appreciation. Indeed, as the editor and garden columnist Allen Lacy points out in the foreword to his 1990 anthology *The American Gardener,* garden writing as a distinct genre all but disappears (except in the commercial magazines and government bulletins) until its resurgence in the 1980s (12). By focusing on aesthetic appreciation, writers for such commercial magazines as *House and Garden,* the most significant periodical in terms of subscribers and dollars spent, trained an expanding middle class of readers to understand home gardening as private cultivation.

Alternative ways to think about gardening as more than a hobby or as participating in richly historical landscapes persisted among writers who identified their gardening not only as private cultivation but also as a political activity—even a radically democratic basis for a more just way of life. These included long-lived Progressive Era writers such as Bailey and his intellectual descendants, Scott and Helen Nearing (whose work is the subject of the next chapter).

Depression-era debates over agriculture focused on the fate of small farms and farmers in the context of the severe crises in American agriculture that Bailey had surveyed. The dust bowl is often evoked as synonymous with a total environmental and social collapse for American farming, a symbol as well as a source of moral lessons.[12] The Dirty Thirties became a turning point, accelerating demographic transition and application of economies of scale, machinery, and agrochemicals, which altered what the terms *farm* and *garden* meant to Americans. Farms, once associated with family homesteads and scaled to an agrarian land policy for settling Euro-Americans west of the Alleghenies, entered an era of accelerated industrialization in the ensuing decade.

Gardens, kept at the edges of crop fields as provision plots by black tenant farmers in the South, as potagers by midwestern farming women, and as front-yard subsistence patches by city dwellers across the country, succumbed to the cultural onslaught of the postwar lawn's homogenous turf. Policies increased the scale of farming, continuing trends pioneered by the so-called suitcase farmers and their fleets of tractors running twenty-four hours a day that broke the last miles of prairie sod. As many environmental and agricultural historians have shown, federal policies, initially intended to keep more rural people on the land—and out of cities—aided industrialization of farming by favoring the big, "efficient" operators and pushing the small, "inefficient," and tenant farmers into industrial jobs or wage work. What would gardening come to mean in the face of the industrialization of American farms?

In the second half of the twentieth century, Americans came to spend millions of dollars on lawns and continued to dose their turf with twelve times the volume of chemicals used by commercial farms, including those that Rachel Carson describes as biocides in her 1962 classic *Silent Spring* (8). How did national culture shift from idealizing backyard food production to marginalizing small-scale, diversified, and localized market agriculture? Sarah T. Phillips's investigation of the cultural and political legacy of the industrialization of American agriculture and her account of the political role of U.S. federal agencies are usefully extended by analyzing the work of the economist Ralph Borsodi. Phillips demonstrates how New Deal policy shifted during World War II away from agrarian liberalism that aimed to help poor farmers keep their farms toward "industrial liberalism" (199, 231–37), which embraced federal involvement in "helping farmers to get big or get out" (200). Borsodi's writing on gardens as domestic economy inadvertently marginalized subsistence agriculture and smaller operations, which thus became more susceptible to industrialized after the depression.[13]

In his 1933 book *Flight from the City,* Borsodi advocates a homesteading scheme based on subsistence family gardens rather than redeveloping market farms. Emphasizing garden production for use, he updates the agrarian ideology of earlier homesteading manuals published in the United States, effectively helping to make suburban development (which, as it threatened country life, undermined an agrarian republican ideal) safe for democracy.[14] In particular, Borsodi imagined gardening as the optimum scale of agriculture for revitalizing the American economy and domesticating radicalized unemployed city workers. He describes gardening within the context of domestic production, which he defines as work in and around the home to furnish basic necessities, including shelter, food, and clothing, that are consumed at the point of production. A critical view on the limitations of Borsodi's ideas becomes particularly valuable given the rise of local eating (the term *locavore* entered the English language in 2007) as a motive for contemporary edible gardens.

Borsodi reengineered domestic production for suburban homesteads and imagined an expanded, though unpaid, role for American women as consumer-producers. His moment of discovery took place in the kitchen one night after returning from his office job in the city: "With the Borsodi family, it was the canning of tomatoes which led to the discovery of domestic production. Out of that discovery came not only an entirely new theory of living; it led to my writing several books dealing with various phases of the discovery" (10). An economist by training, Borsodi argued that domestic production culls the benefits of industrialization, particularly its application of machinery, without its increased costs of distribution, its enslavement of workers to repetitive operations, and its shoddy mass-produced goods. Borsodi further advocated using machines to reduce drudgery, even those home appliances that he admitted were luxuries at the time: power saws, drills, pressure cookers, mixers, grain mills, washing machines. His endorsement of machines as a means to self-sufficient comfort anticipated postwar advertisements of home appliances as well as subsequent critiques of consumer culture that emphasize small-scale technology and decentralized production to remedy the social harms of industrial capitalism. Borsodi's work proposes, in effect, a consumer rather than a political revolution.[15] Finally, the domestic production that Borsodi had in mind depended on depressed land values for mortgaged farmland near cities.[16]

That a self-sufficient homestead might drive economic and political development was not a new idea in 1933; however, since Frederick Jackson Turner touted the closing of the American frontier, the shape and meaning of homesteading had been repeatedly reimagined, the frontier internalized or reinterpreted in economic or political rather than spatial terms. Borsodi's homesteads on former farmland were at the edge of an industrial society in crisis in more ways than one. He held up his own experiments in Flushing, New York, and the Homestead Units of Dayton, Ohio, as the next frontier of America's development: "a new frontier" to which "the enterprising, industrious, and ambitious families shipwrecked in some way by the depression can migrate, just as in all the great depressions of the past century, they migrated from the industrial East to settle on the old frontier" (161). Facing rural and urban industrial crisis, a new suburban frontier would be resettled to recover from the economic depression and avoid social revolution by suburbanizing waves of ambitious migrant workers. The Homestead Units in Dayton were to be created by a local charitable foundation, which purchased failed farms in the suburbs and subdivided them into family plots of roughly three and half acres, space enough for a self-built home and garden.

The political value of a new round of government-subsidized homesteading was clear to Borsodi: it settled the potentially volatile energy of unemployed industrial workers, workers "[who] are apt to become social charges, to become utterly demoralized by public charity, and in the end not only to loathe but to become revolters against a social system which subjects them to such treatment" (122). But the unemployed factory workers were not to become farmers, a possibility foreclosed by the volatility of crop prices, according to Borsodi. Instead of the industrial-scale commodity-crop farming, with federal price support and expanded industrial employment that accompanied the New Deal and the war to come, Borsodi envisioned a decentralized network of homesteads, producing in each household garden the full variety of produce necessary to feed its inhabitants. This newly invented frontier in an old farming landscape was thoroughly domesticated and domesticating. It settled politically revolutionary tendencies and secured economic independence—with a small initial investment of charity and the unpaid work of Mrs. Borsodi's counterparts in the gardens and kitchens.

The political implications of Borsodi's homesteading scheme become clearer when we examine the plans for the Homestead Units, reproduced in the postlude of *Flight from the City*. The plan for the first fifty units echoes the architectural designs of Ebenezer Howard's garden cities. Howard's critique of British industrial cities, *Garden Cities of To-morrow* (1902), called for regional planning to balance the costs and benefits of city and country. His ideas directly shaped the planning of American towns such as Greendale, Maryland, and influenced later regionalists such as Lewis Mumford. The most immediate difference in Borsodi's plan was the sporadic, decentralized distribution of the homesteads compared to Howard's concentrated production in industrial and agricultural zones and zoned residential space at a community level.

Decentralizing production into the nuclei of family domestic spaces—that is, into individual homesteads or home/gardens on a subdivided former farm—predicts the decentralized pattern of suburban settlement in postwar America. Social relations were also implied by the geography of suburban development in the decades following Borsodi's prophetic blueprints. In his plan, the family unit is the nucleus of industry, agriculture, and dwelling; community cooperation occurs between the cellular homesteads, or detached family houses surrounded by gardens. Thus these two spatial maps also represent cognitive mappings and political ideals. Borsodi relocated agrarian ideology in a domestic, suburban sphere with gendered divisions of labor in the home and garden. More important, he sketched a decentralized pattern of suburban settlement with a diminished role for public municipal authority compared to garden city design.

The Jeffersonian idea that individual American families on working farms formed the backbone of an independent citizenry transformed into a modern, appliance-soothed suburban house-and-garden dwelling. Reduced from forty acres to three and a half, Borsodi's version of subsistence homesteading provided a blueprint for suburban development in a country threatened in the 1930s by economic collapse and political revolution. All that remained was to retool the home-garden into a unit of consumption for manufactured goods rather than a base of subsistence production.

Borsodi underestimated the work of renewing a homesteading way of life on the frontiers of cities in economic decline; his subsistence scheme required independent income as well as much unpaid work. In particular, his calculations of domestic production inconsistently accounted for his wife's labor (Mrs. Borsodi's memorable canned tomatoes) (12). His construction of her work as mysteriously incalculable had the practical result that Borsodi admitted to it a lower wage value, though perhaps not as low as contemporary wages for women canning in U.S. factories. He also claimed that the marginal nature of garden and kitchen work made it difficult to determine labor costs: "Even if we had kept track of all the odd times during which we had worked in the garden, that would have helped little without a record of the time put into caring for the single row of tomato plants" and "It proved equally difficult to determine how much time should be charged for the actual work of canning ... household tasks in addition to canning were often performed at the same time" (11). He did reckon, however, the income he lost in order to write as one of the initial "costs" of his homestead experiment (100). Economic undervaluation of Mrs. Borsodi's gardening reveals a psychological overinvestment in her presence in the kitchen and garden.

The "Beautility" of Victory Gardens and the Politics of Wartime Agriculture

Images of the victory garden during the Second World War represent a symbolic resolution to earlier debates over the meaning of small-scale cultivation. National debates over agriculture and land policy during the depression, the specter of wartime hunger, and a dream of postwar prosperity all pushed gardening to the cultural foreground before the national lawn took root. The push to assimilate food gardens for war ends, represented in victory garden propaganda, had the unintended effect of marginalizing those gardens after the war's end. Instead of pleasure, leisure, or independence, food gardens connoted sacrifice, loss, and duty.

Victory garden propaganda also overwrote imported ethnic gardening traditions in American cities with a single national narrative of mass mobilization. In her transatlantic history of the movement, *Cultivating Victory* (2013), Cecilia Gowdy-Wygant claims that their "cultivation was a tool of identity, but one of the values espoused by white, patriarchic American society" (58). The postwar suburban lawn would represent the next step in a national narrative of middle-class mobility and the arrival of the "affluent society," which reshaped postagrarian landscapes and resorted neighborhoods by race and class.[17] In the postwar era, cities reclaimed victory gardens in many places for urban development and parks, while commercial magazines advocated suburban lawns that beautified and celebrated a sanitized, domesticated consumer culture. Finally, as American rural landscapes gradually depopulated and a smaller number of farms scaled up to industrial operations of thousands of acres, the general vocabulary for discussing gardening as agriculture also changed in the popular press. Prewar and wartime discourse focused on domestic subsistence economy, whereas commercial garden writing after the war focused on private cultivation.

Smaller periodicals such as *Organic Gardening* and reprinted homesteading guides continued to circulate after the war, just as thousands of farmers and country people continued to grow and can vegetables for subsistence. But key phrases such as "country life," "food scarcity and security," and "domestic production for use" faded from common use in popular magazines, replaced by reanimated aesthetic and commercial criteria. Beauty and good taste subsumed utility, flowering borders and manicured turf edged out vegetables in the glossy pages of middlebrow lifestyle magazines such as *Home Beautiful* and *House and Garden*. Both had printed lengthy guides to vegetable gardening and designing victory gardens during the war; after the war, they devoted pages to home appliances and tours of gardens of affluent families. The rougher work of subsistence all but disappeared. Mainstream gardening periodicals, in the inadvertently suggestive classification used by the advertising industry, had become "shelter magazines."[18]

Victory garden propaganda is perhaps the best-known representation of gardens made for subsistence in extreme conditions. Images of such gardens still seem insulated from analysis by their patriotic aura. Yet struggles over small-scale agriculture and the formalism of ornamental garden manuals also cropped up in war garden imagery, like unwittingly sown vegetables, so-called volunteers. The rhetoric of victory gardens, and the politics of agriculture more broadly, provide a necessary context for evaluating the significance of postwar garden writing. Striking text and images produced by federal agencies, often by otherwise unemployed artists and writers, claimed beauty and utility in the interest of a nationalist agenda. Victory gardens also displaced earlier notions of public gardening as charity or civic provision for public welfare through land redistribution, as in the many relief garden, vacant lot, and school garden programs that had multiplied since the economic depression of the 1890s. In the wartime invocation of national emergency, government-sponsored garden writing and gardening magazines exhorted a mass audience—and especially women—to join big agriculture at the "food front."[19]

Victory gardens, so named after the First World War, are claimed to have contributed significantly to household diets during the Second World War. In his address to Congress in November 1943, President Roosevelt thanked victory gardeners for producing eight million tons of food in the past year while noting that farmers had broken production records despite war shortages in machinery and fertilizer (67). Victory garden propaganda echoed back-to-the-land manuals, but where the latter had suggested a small farm might be a stopgap for feeding a family in hard times, popular magazines presented victory gardening as a supplement to feeding families otherwise fully employed in a wartime economic expansion. Representations of victory gardens appealed to patriotism, ideals such as self-reliance, and nostalgia for country living. By staving off hunger, these gardens also decreased need for rationed ingredients on the home front.[20] Kitchen gardens would supplement the domestic diet, allowing farmers to sell crops to allies or the federal government to supply its soldiers; in this sense, victory gardeners did join larger-scale farmers on the war's food front. Indeed, consumer product propaganda during the war represented the food front rather literally. One Dairy Products Association advertisement published in a popular gardening magazine at the time displays ranks of milk bottles, cheese wheels, tomatoes, eggs, and ice cream cones marching alongside a plucky GI in fatigues.

In the context of shortages sharpened by food rationing, the popular press conscripted gardens to serve national interests. Middle-class lifestyle magazines such as *House and Garden* incorporated patriotism in their editorial pages. Designs for victory gardens temporarily displaced tours of the flowering hedgerows in posh private gardens. The longest-serving editor of *House and Garden,* Richardson Wright, led the charge to patriotic cultivation. Ideas of gardening as a route to private happiness, moral goodness, and even justice in a universal sense of connectedness to nature and people across differences in a democratic culture, which may have conflicted with wartime needs for increased industrial production, were reinterpreted as one national purpose: victory gardening. Such gardens contributed to national security and secured the gardener's national identity.

Of particular interest is the way in which Wright and others finessed the question of what to do with flowers and ornamental borders. How were gardeners to interpret such perennial beauty? In 1943, Wright went so far as to claim that the spiritual or psychological benefits of horticultural beauty indirectly served the war effort by raising morale. *Gardens for Victory,* a practical guide to victory gardening published during the war, coined a neologism for this combination of purposes: "beautility" (Putnam and Cosper 36). Though an awkward term, *beautility* captures a moral-aesthetic complex, the way victory garden propaganda cultivated military support through the act of gardening.

Gardens for Victory proposed garden designs for a homeowner, describing a detached suburban house complete with garage and hedgerows that belonged as much to a dream of postwar class mobility as it did to patriotism. By 1945, popular gardening magazines turned increasingly to dreams of living in such homes. In his editorial "Food for Peace," Wright suggests that victory gardeners might continue their patriotic work in another war to end war: "Wherever you look and wherever good land is available, Victory Gardeners are at it again, growing food that will help end war" (52). "Food for Peace" linked gardening to postwar nationalism in domestic terms: "This war, as never a war before, has given our country a responsible position in the family of nations. . . . For those of us who till the land, our contribution is food given indirectly. We become self-supporting so that the surplus of our farms can feed people of other lands until . . . they, too, can become responsible, contributing members of the family of nations" (52).

Wright's victory garden, here imagined continuing into the postwar era, justified America as titular head of a "family of nations." Gardens joined an appeal to utility and beauty for patriotic purposes with a vision of middle-class suburban living. Wright extended the metaphor overseas, so that the newly ordered American garden set the template for world order. This hyperbolic riff on the victory garden borrows the paternal language used by American politicians and journalists since at least the time of President McKinley, whose speech in February 1899 reflecting on the invasion of the Philippines epitomizes paternalistic imperialism. At a dinner at the Home Market Club in Boston, McKinley spoke of "the guiding hand and the liberalizing influences" of Americans who were violently suppressing Filipino revolutionaries fighting for independence, and promised that "neither their aspirations nor ours can be realized until our authority is acknowledged and unquestioned" (191–92). Wright's flight of rhetorical fancy echoes this paternalistic view of American international influence and departs from the congenial, ironic tone of his many essays on gardening and travel. It further demonstrates the extent to which the war had enlisted his magazine in its efforts.

As a place and metaphor, the victory garden combined the work of culture and subsistence, transmuting different forms of gardening into one patriotic, national ideal. The idea of the victory garden exerted a centripetal force on the meaning of gardens in wartime America, asserting the primacy of national purpose, where previously gardeners had dug as leisure pastime, as cultural survival, for beauty, or merely to eat. Many postwar garden writers, including those we will meet later in this book, reinstated a sharper aesthetic division between utility versus beauty, a distinction deliberately erased in victory garden rhetoric.

In practical terms, the debate over whether farming would continue to be understood as a way of life at the root of national politics or as a business on an industrial scale was resolved symbolically by the joining of politics, agriculture, and aesthetics in victory gardens. Postwar farming was structured as an industrial business by the political system, and the dominant version of gardening in the mass-market press would henceforth emphasize verdant places of leisure and beauty. In extreme forms, commercial garden writing advocated what Christopher Grampp has described as the "domestication" of home gardens as extensions of the suburban house, making yards outdoor rooms to be outfitted with leisure goods. In *From Yard to Garden* (2008), Grampp follows the design of gardens as outdoor extensions of a family house as it developed rapidly in the Mediterranean climate of California, in bungalow architecture, and with the so-called California-style of gardening (158). Postwar consumer culture turned earlier ideals of garden-based subsistence inside out, rendering domesticated yards blank slates to be remade annually with the newest gadgets.

Removing kitchen gardens opened more space for lawns and allowed the narrow range of garden design (turf with foundation plantings and sparse trees, with no fences or hedges) that came to dominate suburban landscapes in the United States well into the twenty-first century. Perhaps without recognizing it, Americans were invited to subtract agriculture from the culture of gardening with the achievement of victory and establishment of the great postwar lawn, becoming what the geographer Paul Robbins has called "lawn people." It was as political an act as the enlistment of home gardens for food during the world wars, but not without its critics: a vocal minority of dissenting, eclectic, and sometimes radical American writers on gardening had other ideas. In the remaining chapters, I will attend to their voices and landscapes over the last century and unearth in their work the uneven traditions of gardening for democratic culture and the ideals of justice.

2 POSTWAR GARDEN WRITING, LITERARY CULTIVATION, AND ENVIRONMENTALISM

In the postwar United States, subsistence and botanically diverse gardens, with their complex arrangements of plants, yielded to an almost puritanical substitution of chemically dependent, ill-adapted turf. The fenceless suburban lawn had been opposed by garden writers at the opening of the twentieth century, communicating from their privileged perspectives behind the high hedgerows of established country estates and older suburbs. Late Victorian and Progressive Era garden writers such as Alice Morse Earle, Neltje Blanchan, and George Washington Cable all agreed that the trend toward fenceless front-yard lawns sacrificed much of value—including fragrant flowering borders as well as the privacy and freedom afforded by a tall hedge. Cable referred in 1914 to "American gardens' excessive openness," noting that "except among the rich, [they] have become American by ceasing to be gardens" (61). By the end of the twentieth century, writers were launching ecological and cultural criticisms of the postwar suburban lawn. A few began linking food gardens implicitly with environmental justice. The concept that all people have a right to healthful places to live, work, and play, free from unfair exposure to risk, drew attention to the actual inequality of living conditions and social geography of American society, despite its ideal of openness and democratic equality represented in the design of postwar suburban landscapes.

Twentieth-century writers in the United States aligned gardens with democratic, moral, and environmental commitments, converging on fair access to amenities in urban areas as the environmental justice movement emerged near century's end. So when, in the first decade of the twenty-first century, garden activists called for "Food Not Lawns," planting their front yards with hardy kiwi vines, heirloom ghost eggplant, and dwarf apple trees instead of fescue and bluegrass, they drew on a half-forgotten and scattered garden literature. This chapter tells the story of two more or less forgotten postwar writers whose work helps define the range of aesthetic and political expression in U.S. garden writing. Each represents a buried tradition of values other than those dominated by the stunning growth of consumer culture, which by the end of the twentieth century had made lawn care—if not gardening—a business of some $40 billion per year.

The chronic and acute environmental health problems resulting from this intensive pesticide- and fertilizer-heavy turf culture have been dissected by environmental historians, historians of medicine, and critical geographers, notably in Paul Robbins's *Lawn People* (2007), a history of lawn culture. These range from loss of hearing and cases of poisoning among professional landscapers, many of whom are recent immigrants, to the overuse of chemical fertilizers by earnest home gardeners in early-spring lawn applications, which wash into surface waters and cause algae blooms and oxygen shortages. American gardeners continue to use relatively inexpensive and widely available garden pesticides, including 2,4-D and other lethal chemicals, despite risks to human health identified in Rachel Carson's *Silent Spring* in 1962. The application of lawn-related pesticides amounts to a legal exposure of incremental toxins to millions of citizens.[1] Tens of millions of tons of pesticides are applied annually to U.S. lawns, or by one 1996 estimate, five to seven pounds per lawn each year.[2]

Many of these chemicals have known mutagenic and carcinogenic properties. But the vast majority of pesticides licensed for nonfood application on lawns are not tested by the U.S. Food and Drug Administration or the Environmental Protection Agency for chronic health effects and neural toxicity.[3] Therefore, the full cost of chemical lawn applications is unknown. These costs are not rooted in an American gardening ideal but rather in the lucrative marketing of lawn care to postwar suburbanites. In the decades since the Second World War, U.S. homeowners became what Robbins has called "turfgrass subjects" (129), the product of political and economic forces of socialization. The culture's unspoken mantra reversed George Washington Cable's formulation and realized his feared scenario: lawns replaced gardens, particularly as the preferred landscape of an expanding middle class.

Not everyone shared the enthusiasm for a national greensward. Rural people, recent immigrants, more ecologically minded organic gardeners, and some urban elites preserved earlier notions of gardens as private, intricate spaces notable for their enclosure of biocultural diversity. In thousands of eclectic nonconformist landscapes, American gardeners perpetuated intergenerational traditions threaded invisibly through seedways across continents and oceans, cultivating epazote, hawthorn, mountain ash, okra, yams, Oaxacan dent corn, and custard apples. Patricia Klindienst has explored the ways in which ethnic American gardeners made homes for themselves in her eloquent book *The Earth Knows My Name* (2006), based on interviews conducted countrywide. Particularly for political exiles and recent immigrants—whether from Mussolini's Italy or Indira Gandhi's India—gardens became places "where they have claimed, or reclaimed, a just and good relationship to the earth" (242). This widespread, quietly acknowledged linking of gardening with justice predates the articulation of environmental justice by U.S.-based activists in the 1980s and maps cultural routes among ethnic gardeners that exceed national boundaries. Klindienst's book effectively provides a voice for multitudes of U.S. citizens who experienced gardening as linked with environmental justice. Perhaps because they also wrote in genres other than those favored by high literary culture (e.g., the novel, lyrical poetry), the postwar garden writers examined in this chapter also delved into protoenvironmental justice themes.

As late as 1970, Katharine S. White described American garden writing with a note of despair over its literary quality. Except for "a few shining exceptions," she wrote in her *New Yorker* magazine column, "American 'garden writers' are a drab lot" (333). According to White, garden writing in Britain had been a serious literary genre since the Elizabethan herbalists, whereas American writers of her day produced manuals and commercial books of little serious interest. White's judgment echoed a persistent anxiety in American culture, one repeated by later U.S. garden writing enthusiasts who also sorted the agricultural from the cultural in collections of garden writing and looked to Europe for aesthetic inspiration. A few homesteading tracts and manifestos published between the 1930s and 1950s (notably those by Ralph Borsodi and Scott and Helen Nearing) focused on gardening as a necessity in hard times or as a means to reconnect with nature. Yet later anthologies of American garden writing usually skip over their work and the period, perhaps sharing White's assessment that the "drab lot" of midcentury U.S. garden writers did not merit their time.

White served as the fiction editor of the *New Yorker* from 1925 until her retirement in 1958, after which she wrote a seasonal column reviewing garden books until her death in 1970. As editor, she promoted young talent and published many well-known American modernist writers, including her husband, E. B. White. The literary judgment wielded as an editor clearly extended to garden writing—though she claimed to read garden catalogs the way many read novels. Yet clearly for White, not all writing about gardens by Americans should count as American garden writing. Ignoring almanac writers, coffee-table editors, and glossy magazine publicists on literary grounds obscures a great deal of interesting cultural work and environmental history. Gardening in America, after crop agriculture and cities, is the most substantial, widespread, and everyday way in which people have transformed their environments, "mixing their labour with the earth," as Raymond Williams wrote in "Ideas of Nature" in 1980 (76). Such a fundamental activity leaves traces at the levels of literature and culture. Gardens are where many everyday concerns and mediations of environment take place, and U.S. garden writers were among the first to voice concerns over the use of pesticides and to express ambivalence toward profit-driven consumerism even while engaging in the large-scale introduction of new species for commercial as well aesthetic purposes.

Such concerns had appeared in the work of Liberty Hyde Bailey, possibly one of the "shining exceptions" to which White referred. But the economic, political, and environmental crises at the heart of rural life in 1930s America and the rapid changes that followed the Second World War (explored further in chapter 6) jeopardized the meaning of small-scale cultivation. The postwar country was reshaped by dramatic urbanization and accelerating suburbanization, with heavy industry and, later, service economy jobs supplanting farming as the livelihood of most Americans. Small-scale food gardens and the heroic figure of the individual gardener temporarily found a leading role in national culture with the victory garden mobilization of the 1940s. Reading the Nearings alongside White and the mainstream garden magazine editor Richardson Wright allows us to recognize a spectrum of concerns broader than each writer's interests. These include garden design, modernist literary style in postwar culture, the toxic health effects of industrial food, and the destructive environmental effects of American lifestyles.[4]

If a progressive class of refined gardeners had produced an abiding vision of gardens as central to democratic culture over the preceding two centuries, America in the decades before White's categorical dismissal was fast on its way to becoming what the historian Liz Cohen calls the "consumer's republic." In a society increasingly fragmented into groups defined by consumer preferences rather than civic duties, perhaps literary taste became even more important for an editor in White's position. Yet a variety of interpretations of gardening—most having little to do with high culture or literary taste—competed for readers in the mid-twentieth century. Photo journals of lavish garden tours, magazine editorial columns, personal essays, government handbooks, and horticulture and design manuals imagined gardens in ways as diverse as their makers. What was the purpose of literary cultivation if not to identify which among these prescriptions for gardening continued to be worthwhile? How would cultivation, understood as a deliberate self-aware fashioning of values, and cultivation as working land come to serve a culture devoted to consumption, which masked environmental connections? My method in pursuing these questions is more historical and discursive than White's; rather than policing a national genre of reflective "garden writing," I aim to link many forms of writing on gardening. Doing so will reveal the tensions and recover marginalized voices that anticipated ideas of environmental justice.

White's use of scare quotes around "garden writers" to denote the producers of a genre can be put to the service of this fuller cultural history. Such a history of garden writing must include marginalized writers and the broader sweep of the celebrated garden authorities of the eighteenth and nineteenth centuries. This included presidents and park designers but depended on a mass audience of anonymous gardeners whose practices in their vernacular landscapes could diverge wildly from generic prescriptions. Yet collectively these writers represented six or seven generations of democratic, public-spirited seed traders, gardeners, landscape architects, editors, and scientists. This intergenerational group shared a deep sense of the redeeming power of human actions in coordination with nature, so it makes sense to call them "progressive gardeners," borrowing the designation that Bailey used for those horticulturists and plain dirt gardeners who adapted the best scientific methods for cultivating plants. That hopeful emphasis persisted in the old Unitarian slogan that White borrowed as the title of her column, "Onward and Upward."

Postwar garden writing also reclaimed some of the meanings of gardening that were overshadowed by mass mobilization for food production during the war and the nationalism of the victory garden.[5] In the immediate postwar era, commercial garden magazines reverted to a division between beauty and utility, returning to an aesthetic appreciation of flower gardens and sophisticated estate designs. The authors I consider in most detail here represent more than one alternative to commercial garden writing; they exemplify parallel literary formations, discrete modes of practicing literature, including styles of writing, frames of reference, and attitudes toward politics in the broadest sense. Despite obvious biographical similarities, historical coincidence, and geographical proximity (at one point, the Nearings and White gardened on farms in coastal Maine within a dozen miles of each other), they developed distinctly radical and liberal-reformist ideas that deserve to be considered together as part of a discourse of gardening in the United States.

White made gardening a marker of cultivation and taste in her *New Yorker* columns. For her, the literary value of garden writing continued an Anglo-American tradition of gardening as an aesthetic rather than subsistence activity. Consequently, she did not view as exemplary other types of writing about gardens, whether those merely useful for raising vegetables or overtly political. Scott and Helen Nearing's work has been long ignored by literary critics, although the historians Rebecca Gould and David Shi have given them their due in separate histories of homesteading, secularized spirituality, and simple living in America.[6] Both have traced the Nearings' influence on the back-to-the-land movement of the 1970s. The Nearings' writing was more concerned with subsistence than literary cultivation and holds gardening as key to a good life. It proposed a radical political ecology reconnecting land, power, and working people, rather than as garden writing as a narrower literary genre. In *Living the Good Life* (1954), the couple presented a plan to achieve a more just and ecologically sound social order in which gardening—in particular home food production using organic methods, as articulated by Albert Howard and, later, J. I. Rodale—takes over the small-scale subsistence role laid out by Ralph Borsodi that had once been associated with farming.

Gardening as Revolutionary Practice in the Nearings' Good Life

Scott Nearing, like Ralph Borsodi, was an economist by training and a Progressive Era reformer, although he became a self-professed radical by inclination. As a young economics professor, Nearing spent summers living with his family in Arden, Delaware, an intentional community founded in 1900 that was inspired by the arts and crafts movement. Many Arden residents designed and built their own homes, raised gardens for seasonal food, and dedicated time to community arts. The experience of being repeatedly ousted from university posts, first from the University of Pennsylvania in 1915 after advocating for a child labor law opposed by members of the university's governing board and then from Toledo's College of Arts and Sciences for his vocal pacifism in 1917, radicalized Nearing. After federal agents raided his home in Toledo, he moved to New York City and began lecturing at the socialist Rand School. By the 1930s, he had been blacklisted by his publishers, even though his economics textbook had become the standard assigned at many universities, and he was blocked from the progressive lecture circuit because of his unorthodox socialism and pacifism. As he writes in his autobiography, *The Making of a Radical,* "I chose homesteading as a way of life under United States right wing pressures in the 1930s" (210). Scott and his partner and eventual second wife, Helen Knothe, bought an abandoned farm near Jamaica, Vermont, in 1932 and soon left New York for good.

The farmstead became the setting for their self-published accounts of subsistence living. They wrote these accounts together in the first-person plural with few attempts, except in a few humorous asides, to distinguish their individual perspectives on shared experiences. In *Living the Good Life,* the Nearings' garden took center stage as the metabolic engine for their alternative "use economy," which they acknowledged adapted some of Borsodi's ideas about domestic production and appealed to American self-reliance. (David Shi's 1986 history of simple-living movements in the

United States, *The Simple Life,* places the Nearings in a longer tradition of simple-living and utopian communities [255–57], as did Paul Goodman in his 1970 introduction to the revised edition of *Living the Good Life.*) Simultaneously working from a tradition of homesteading literature and reversing its growing market orientation, the Nearings presented a functional guide to a rural domestic economy not, as Borsodi had imagined, as satellites of an urban center dependent on manufactured home appliances, but rooted in craft skills and building the soil.

Although the Nearings' back-to-nature model helped inspire 1970s environmentalism, their political critique of American capitalist culture was resoundingly Old Left.[7] They celebrated the physical labor of gardening and praised how it brought them back into touch with nature in a material as well as a spiritual sense. Their first guide presented an idiosyncratic critique of the status quo in terms of the city and its exploitative economy, which they transmitted as a guide for living that included much practical wisdom: organic gardening techniques, directions for building homes out of local stone, and a detailed vegetarian diet. *Living the Good Life* belongs in a sequence of philosophical and political writing on gardens that stretches from Virgil's *Georgics* through William Cobbett's 1854 treatise *An American Gardener,* books that also contemplated the significance of rootedness and human connections to nonhuman nature and (in Cobbett's case) the perspective of political exile.

Scott and Helen Nearing's move to Vermont in 1933 was precipitated by the cultural disintegration of left-leaning publication and speaking venues amid the wider economic crisis of the Great Depression. After decades of agricultural, economic, and ecological collapse, the Nearings called their gardening "bread labor," borrowing the term from Leo Tolstoy. Like Tolstoy, agrarian socialists, and the British reformer William Morris, the Nearings looked to country ways for residual values to oppose the drift of what they viewed as a destructive modern way of life. In *Living the Good Life,* they modified the values circulating around the American farm for practical and theoretical reasons. Returning to a farming landscape, they nonetheless broke from local practices. They refused to raise animals for meat or sale and grew only enough for their needs, combining progressive politics with hard physical work. Nor were they sentimental about farming as a way of life. They rejected the common practice of keeping animals, either to make plowing easier or food more abundant, on the grounds that husbandry was cruel and uneconomical. More important, they rejected farming as a means to participate in the wider consumer economy. They deplored their farming neighbors' reliance on packaged goods, which they described as "dependence" on "poisoned, processed, and chemicalized foods" (131). Rather than expanding their cultivation to match demand for a cash crop, the Nearings limited themselves to subsistence gardens, food preservation, and a rekindled sugar maple operation to fund necessities they could neither produce themselves nor obtain through local trade, notably fuel and a pickup truck for hauling building stone to terrace uneven terrain for their gardens.

Despite naming their homesteads in Vermont and Maine "Forest Farm," the Nearings invariably wrote about their work as gardening rather than farming. In scale, their gardens could just as easily have been called truck farms, were they producing for a local market. The inherited scale of agriculture in northern New England on three to five acres was (and remains) an accommodation to local conditions: a northern climate, rocky and often eroded soil, and only a distant substantial market for perishable commodities. These made market farming a marginal enterprise at best.

Such conditions could, however, provide a livelihood for political exiles, such as strident leftists, during the decades of intense red-baiting after the Second World War. In the Nearings' work, the political debate over American agriculture, which had focused on the small farmer as an endangered producer and farming as a declining way of life, developed into an outsider's socioenvironmentalist critique of the links between an industrial food system, threats to human health, environmental degradation, and social injustice. Their conscious, if anxious, retreat from American politics is also suggested in their choice of the term *garden* over *farm*.[8] Small-farming manuals, published regularly from the 1890s to the present, provide only the generic format for their guides; the Nearings' critique of American industrial consumer society signaled an estrangement from new industrialized farming and a turn to organic gardening and community living.

While critical of many of their neighbors, the couple documented the farm's previous owners with a degree of sympathy. In 1954, they presented a bleak picture of rural Vermont as a culture in decline, and though they often praised the value of community, tensions become apparent when they tried to reach beyond their own boundaries. Outside the simple calculus of their use economy, living in rural Vermont appeared more complicated. The Nearings recognized that they "did not conform to the folkways" of the Green Mountains and "were not native sons," though they homesteaded there for twenty years (166). Neighbors were at turns "amused," "baffled ... or annoyed" by their habits: "Perhaps the most consistent and emphatic disapproval was directed against our diet. We could more easily have been accepted if we had eaten in the approved way. We ate from wooden bowls, with chopsticks ... we ate food raw that, according to Vermont practices, should have been cooked, and we cooked weeds and outlandish things that never should be eaten at all. That we ate no meat was in itself strange. ... We simply failed to live up to the accepted Vermont pattern" (167).

The story of their support for a local community center and discussion group underlines how their politics and outsider status put them at odds with other rural people. According to the Nearings, a planned community house in the valley failed because the locals were split over whether to serve alcohol at weekend dances. (The Nearings were teetotalers, but many locals appreciated a drink with a dance.) Likewise, the Nearings' reading and discussion group diminished in size after 1939. They disagreed with the militarism of their neighbors, who in turn saw them as "too radical" in their opposition to the war. The Nearings rationalized that this political impasse to living in a community was "not surprising in view of the fact that the Vermonters, Republicans almost to a man, looked upon Democrats as way to the left" (177). As unorthodox, nonparty communists, the Nearings also opted out of the popular front, holding to their uncompromising pacifism.

It was the political context of anticommunism, after all, that had sent the Nearings into what they described alternately as self-chosen exile and pleasant life in the country centered on their garden homestead. A garden could produce a world-changing economy of use: "The basis of our consumer economy was the garden. By raising and using garden products ... we were able to provide ourselves with around 80% of our food" (153). They represented gardening as a practical solution to two historical problems: on the one hand, an exploitative exchange economy and its industrial food system; on the other, an erosion of personal virtue. To dramatize these problems, the Nearings compared health-giving whole foods, raised in their Vermont soil, to the specter of "poisonous" food and the market economy of their previous home in New York City. Processed foods represented a synecdoche for the oppressive living conditions they associated with the market and city ways, "the whirlpool of New York life," as opposed to "the tranquility of hills and forests" in the "Vermont wilderness" (159). Theirs was a polemic against the city as the center of commodity distribution, consumption, and waste, expressed in pastoral terms: "Coming from New York City, with its extravagant displays of non-essentials and its extensive wastes of everything from food and capital goods to time and energy, we were surprised and delighted to find how much of the city clutter and waste we could toss overboard. We felt as free, in this respect, as a caged wild bird who finds himself once more on the wing. The demands and requirements which weigh upon city consumers no longer restricted us" (155).

Significant judgments are implied in this chain of pastoral associations: the city, a cage, clutter, waste, "limitations, restrictions, and compulsions," and ultimately the demands and requirements of a market economy. On the other side stand the country, wild birds, efficiency, freedom, and a "household economy based on a maximum of self-sufficiency" (156). A use economy, or self-sufficient rural economy as they elsewhere described their Vermont project, would trade in skills instead of money. They contrasted the moral value of the "ingenuity, skill, patience, and persistence" needed to succeed "in the backwoods" to the pecuniary value of the market: "The store customer, who comes home with a package under his arm has learned nothing, except that a ten dollar bill is a source of power in the market place. The man or woman who has converted material into needed products via tools and skills has matured in the process." Of the backwoods virtues, the Nearings identified "the use of the soil and the production and preparation of food" as the most important (158). What began as a critique of the urban living conditions produced under the American capitalist system ended as a paean to the supreme virtue of organic gardening for cultivating character. This pursuit of personal virtue through organic production stands in tension with their systematic political critique, which otherwise anticipates many of the concerns of the food justice movement in the early twenty-first century.

The Nearings' garden-centered vision also evokes familiar antimodernist oppositions of city and country, metropolis and wilderness, rustic virtue and urban vice. They imagined their gardens as a homestead in a wilderness, rather than as an agricultural landscape in the backwoods of the metropolis, a prime example of the tendency of pastoral literary conventions to overtake historical consciousness. The presence of a market for the Nearings' cash crop—maple syrup in Vermont and blueberries in Maine—reveals this antagonism between city and country, market economy and domestic economy, as rhetorical. The Nearings presented their cash crop as an inevitable, though limited, compromise with the necessity of a market economy. When writing of food gardening, they imagined they could feed an intellectual life free of such compromises.

The doctrinal purity of their philosophy of gardening stands out against the ambiguous rhetorical texture of their writing. In the epigraphs of each chapter the Nearings cite earlier agricultural handbooks with a colloquial smattering of philosophical insight and practical wisdom. Their references stretch back centuries and across the Atlantic, but apart from directly placing themselves in a cultural tradition, they did not foreground any self-consciously literary qualities shared between their "gardening project" and farming literature and pastoral poetry. When they wrote about writing, they contrasted the work of writing and thinking to their bread labor and social exertions: the good life, they maintained, balanced these three types of labor. It first divided them into a hierarchy from most urgent (material work, bread) to most worthy of a life's purpose (social work, teaching). Writing, in their eyes, was a vehicle for teaching a progressive cultural tradition and thus the intellectual work that connected material and social labors—but the subtle weaving of allusions and citations clearly exceeds the simplicity of this instrumental view of writing.

Although it draws on a living tradition of progressive garden writing, *Living the Good Life* also owed much to Progressive Era conservationism, with its horror of waste and utilitarian approach to natural resources. The Nearings' moralizing language of excess, restraint, and virtue perhaps appealed to a later generation of Americans sensitized to environmental and social crises than did their exhortations to return to homesteading. In the counterculture of the 1960s, their coauthored *Living the Good Life* found a new audience. The book gained a wider audience, particularly among young people who flocked to their Vermont homestead. In fact, as the Nearings reported in their sequel *Continuing the Good Life,* they chose to move to a more remote farm in Maine in part because so many visitors were interrupting their work (358). Certainly they recognized that these "thousands of young people" were "seekers" who had rejected the status quo of a "world community that has tolerated war," but the Nearings also saw the majority as "so many unattached, uncommitted, insecure, uncertain human beings" (359). Again and again, they returned to metaphors of rootedness and emphasized how gardening provided a secure domestic economy and gave their lives focus and political meaning.

In this respect, their writing represented a tradition of mixing conservative agrarian tropes and radical democratic hopes, a tradition that reached an apex of literary prestige with the rise of naturalist and realist novelists such as Theodore Dreiser, Upton Sinclair, and John Steinbeck and the brief heyday of proletarian literature, only to be marginalized by the cultural institutions of the postwar era (Denning xvi). As with many features of American literary culture, the road to a broader audience for postwar garden writing led back through New York.

The *New Yorker* Edition of American Garden Writing

Rereading Katharine White's collected *New Yorker* columns, published as *Onward and Upward in the Garden,* in the context of her contemporaries the Nearings highlights the unspoken criteria of the "literary" in American culture and its hierarchical sorting of different forms of writing about gardens. White, the fiction editor at the magazine during its first thirty years, presented gardening as a leisure activity without the earnest political emphasis of the Nearings. Her essays reveal a writer as interested in distinguishing a literature of gardening as in getting her hands dirty. White interpreted garden writing as an Anglo-American tradition plagued by flashy commercial success and perfunctory manuals: "We publish each year an enormous number of books related to the garden, but except for our excellent reference books they are mostly how-to-do-it books offering advice . . . or they are superficial and overexpensive de-lux gift books of one sort or another that are bought for their color illustrations, their huge size, and their display value rather than for their text. Unlike this country, England has since the Elizabethan Age regarded books that deal with horticulture and plants as a true branch of literature ... whose books give pleasure not only to gardeners but to any other reader" (333).

White called for a national audience for garden writing constructed in terms of taste—a literary and aesthetic genre rooted in an imagined European past rather than one contextualized in its contemporary multi-ethnic social history. Good American garden writing, according to White, would give literary pleasure rather than mere horticultural instruction. The anticommercial formal criteria here staked a claim for an autonomous realm of literature, independent of its content, something analogous to the formalist privileging of literary modernism that became dominant in mid-twentieth-century universities and literary magazines with the rise of New Criticism and close reading at the expense of historical, political, or moral interpretation. Yet perhaps because of, rather than in spite of, its self-consciousness as "literary cultivation," White's garden writing also prefigured the strengths and limitations of mainstream environmentalism, which has long been catalyzed by publishing successes (from *Silent Spring* to Elizabeth Kolbert's *Sixth Extinction*). White's garden essays also critiqued the presence of synthetic chemicals in the home garden in the lead up to the magazine's serial publication of Rachel Carson's essays on pesticides.

White modeled a critical perspective that many would find familiar from the precise, light touch of editorial style at the *New Yorker*. Like the magazine that she helped establish as one of the longest-running literary institutions in the United States, White took popular culture seriously, attending to the quirks and drama of seed catalogs as if they were novels or poetry. Her essays freely mix reviews of published books, autobiographical narrative, and cultural commentary. Underlining the importance of the catalog writers, White notes: "They have an audience equal to the most popular novelist's, and a handful of them are stylists of some note" (3). Whether on aesthetic or economic grounds, then, the catalog writers and editors affected culture broadly, and their effect could cut across class and region. White was both a critical reader and an advocate for such writers, whom she said "are my favorite authors and produce my favorite reading matter" (5). Such praise meant more coming from an editor who had published the early writings of James Thurber, John Cheever, Frank O'Hara, John Updike, and Mary McCarthy and built the reputation of the *New Yorker* as a national literary venue, what *Dial* and the *Atlantic* had been in the previous century.

White's democratic enthusiasm for popular culture was accompanied, however, by a rather privileged view of gardening as leisure. She maintained a town-and-country habit and instructed her gardener, Henry Allen, where to plant the bulbs in the fall, as E. B. White recalled in the preface to his wife's posthumous collection of garden essays (xviii). Her tone shared in the ironic modernist sensibility used by other garden writers from the 1920s in Europe and the United States, for example in Vita Sackville-West's arch declarations on her British estate at Sissinghurst, Richardson Wright's send-up of his own garden puttering and bourgeois marriage in *The Gardener's Bed-Book* (1929), or Karl Čapek's parody of himself as an obsessive rare-plant enthusiast in *The Gardener's Year* (published in translation in 1931). At their most critical, these three writers presented gardening as comical and lighthearted: taking pleasure in such writing depended on not taking oneself (or one's gardening) too seriously. Such humor assumed a context of comfortable domesticity, along with a class of readers surrounded by (or longing for) material comfort and, in Wright's case, American suburban affluence.

In contrast to these comic garden writers, White's essays treated the textual gardens of catalogs, handbooks, and picture books more often than her own landscape or plants. She invoked her Maine garden to set the scene and season, though usually only in general terms, for example: "In this cold winter of 1960—cold, that is, in the Maine coastal town from which I write—I have been thinking about the fragrance of flowers . . ." and "By August a flower garden, at least on the coast of eastern Maine, where I live, can be at its best—and at its worst" (103). After these gestures, the essays move on to pithy anecdotes, reviews of books and catalogs, and insights from a wealth of writing on gardens.

White also injected humor into the most personal remembrances. In "War in the Borders, Peace in the Shrubbery," her flowering borders, which appear in rare lush detail, became a stage of mock-heroic battle against the more aggressive plants of her garden in Maine: "The wars of aggression that I thought our private Security Council and its little army of two, armed with spade, fork and trowel, had settled in early spring have started again. The lolloping day lilies have begun to blot out the delicate columbines, the clumps of feathery white achilleas are strangling the far more precious delphiniums, and the phlox itself is at the throats of the lupines and the Canterbury bells. Even the low plants at the front of the borders are making aggressive sorties. The ajuga . . . is one of the worst offenders. Unless I soon repress its insinuating roots, there will be no violets, pansies, or pinks next year" (104). The "War in the Borders" column, which reviewed catalogs and recommended varieties of flowering shrubs and perennials, appeared on September 24, 1960. Its language of borders, "wars of aggression," and logic of containment by "repress[ing] . . . insinuating roots" gently poked at then-current tropes of American military strategy in the Cold War. In not taking her garden too seriously, White encouraged readers to chuckle at the obsessive repression of other menaces to the nation's boundaries. Gardening here represents a metaphorical escape from militarism; for the Nearings, it had offered a practical outlet.

In the same column, White admits that her crowded border "makes me want to say, 'Oh, let it go. Let the plants fight their own battles'" (104). This letting go triggers a nostalgic reminiscence on a childhood game of "Millinery" that she played with friends in her Boston neighborhood. The girls cut pieces of lilac, hawthorn, and forsythia, decorated their straw hats, and then "paraded into the house to show them off to our elders" (106). White claims to prefer flowering shrubbery because it "need[s] less care, less adjudication, less ruthless cutting back than perennials," but its appeal as a peaceful, mazy background of psychological attachments also seems relevant. Gardens, for White it seems, are fundamentally places of restorative connections and perennial blooms, not a means of subsistence or a path to political autonomy as they were for the Nearings.

The reproduction of social classes through distinctions of taste, as the sociologist Pierre Bourdieu illustrated in the case of twentieth-century French society, provides a bridge between White and the Nearings. It would seem on the face of it, with White's preference for flowering shrubs and the Nearings' politically minded food gardening, that no two garden writers could be farther apart in their tastes. However, both White and Scott Nearing wrote of valuing the natural world as a source of moral education and equated conservation of nature with virtue. Both were children of the Progressive Era and its ideas of the social gospel and conservation.⁹

Yet they sought a different nature through different cultural tastes in gardening, and that cannot be separated from their work's acceptance or questioning of the established social order. When set against the postwar American landscape and the economic politics that hastened the vacancy of the farms in Maine purchased by these garden writers, conservation of nature can seem identical with conservation of a social order. Certainly, since Downing and Olmsted, gardens and cultivated nature had been understood as a means to secure class harmony. But the flower gardening of primary interest to White belonged to an upper-middle-class cycle of leisure and work. Home gardens were subject to divisions of labor in terms of gender, class, space (backyards, front yards, suburbs, farms) and time (weekends and seasonal work). In her practice as a writer, White's vision of pursuing the good life in gardens—reflected in the title "Onward and Upward in the Garden"—required a division of labor between an educated class of intellectuals, writing about their farming and gardening, and their hired help. In the Whites' case, the latter was an able gardener and farm worker named Henry Allen, who also cared for E. B. White's livestock on the hobby farm that provided some of the most beloved scenes for his children's book *Charlotte's Web*.

White's collected columns also provide a window into the material culture of leisure-class gardening. The reading of catalogs and garden-writing advice columns plays a role in reproducing tastes, literally in choosing which flowers to plant and whether to plant particular vegetables to eat. At the same time as her work subtly shaped readers' tastes, White satirized gardening as a mass culture phenomenon in the postwar period, noting how it reproduced middle-class values of competition (new and bigger hybrid blooms!) and enthusiasm for chemical pesticides, fertilizers, and new consumer goods (171).

The negotiation of taste in White's writing is more complex than simple class prejudice. In several instances, she used her column to critique garden snobbery, as when she lampooned American flower shows and flower-arranging contests. She parodied one such competition in Florida in the second of a two-part essay published in November 1967. "There were mighty few flora in the arrangements," which instead were dominated by "what are called in flower-show language 'accessories.'" In fact, several of the nineteen classes of competition "allowed only dried or painted material," including wood, metal, and paper. The "Avant-Garde" class awarded the blue ribbon to a "paper-covered cardboard cylinder...painted red and peppered with red sequins stuck on while the paint was wet... [which] held some flowers and artificially curled papyrus." White assessed the red sequined concoction as "Camp, perhaps... but not Avant-Garde" (281). This judgment, representative of her tongue-in-cheek appraisal of other experimental pieces of garden art, suggests traditionalist or even populist horticultural tastes.

White's populism extended to her view of the garden clubs and organizations that had solicited the creation of the flower arrangements she reviewed. Though not a member, she praised the work of garden clubs at the local level: "They plant the public places that need planting, they teach conservation, they befriend wild flowers, they befriend the birds—natural allies of all gardeners.... The little clubs... also provide a democratic meeting place" (271). The state and national associations, however, dictated rules and what she saw as the byzantine procedures of competitions (272). She singled out the National Council of State Garden Clubs for "the heavy hand it lays on American flower arranging." One result was a preference for abstract form over plants as living organisms: "Design is now everything; flowers, as such, don't count. The old idea of bringing flowers into the house and preserving their natural beauty for several days in a bowl of water is forgotten" (283). Invoking flowers' "natural beauty," White asserted what she saw as a more popular taste shared by "women, like me, who belong to no club or society, attend no school of flower arranging, and are simply average householders—country dwellers or office workers or suburbanites." Gesturing to commonplace and popular (if gendered) practices of flower gardening, White claimed "it is easy to believe that half the women of America are putting fresh flowers into vases fairly constantly" (274). This appeal to something like a national sisterhood united by a love of garden flowers echoed the earliest declarations of Louisa Yeomans King and other Progressive Era leaders of the garden club movement.

Elsewhere, White argued that the weedy goldenrod be named America's national flower, imagining the nation represented by a democratic, ubiquitous "weed." Goldenrod had both literary and personal significance for

White, as shown in the most vivid and intimate of reminisces published in her column. She recalled how, as a young teenager vacationing with her family at Lake Chocorua in New Hampshire, they won first prize in a regatta by decorating their boat as the funeral barge in Alfred, Lord Tennyson's poem "The Lady of Shalott":

> My sisters and I began the day by gathering bucketfuls of goldenrod, which my aunt spent the morning stitching onto a sheet in an intricate pattern—no stems or leaves showing, only the plumy golden blooms. . . . My arms were crossed on my breast, the letter to Sir Lancelot was placed in my right hand, and the goldenrod cloth of gold was drawn up over me. . . . Then we circled endlessly in the parade. At one point, the oarsman [her father] removed the letter from my tanned fingers . . . and handed it to one of the judges. . . . Without this message, I fear we would not have been recognized and would not have won first prize. For years, I kept the trophy—a framed photograph of Mount Chocorua with the lake at its foot—knowing all along that it was my aunt's lovingly sewn goldenrod cloth of gold that had won it for us. (189)

In this recollection of wildflowers and Victorian poetry, both materials became a part of the writer's body. The uncertain source of her family's distinction in the "parade" through this more than decorative use of cultivation also asserted culture as a just source of social distinction. This interweaving of literary and horticultural dimensions of gardening exemplifies White's form of cultivation. The threading of poem into real materials, of flowers and words, distinguished White from the crowd. She persisted in this work by transforming the common wildflower through association with the poet (Tennyson), whom White associated most with gardening.

This is not to say that White's garden writing lacked a critical edge. Along with her literary cultivation, tied ambivalently to class-based assumptions about nature, literature, and American culture, came a pointed critique of pesticides, herbicides, and genetic wrangling with plants. At the *New Yorker*, White also played an important role in catalyzing postwar environmentalism, not the least by introducing a broad audience to Rachel Carson's work. She urged publication of the serial version of *Silent Spring*, and in a private letter she told Carson that of all the writing she had published at the magazine, she was proudest of *Silent Spring* (qtd. in Davis

209). Would Carson's publisher have stood up to the chemical industry's threats of litigation had the *New Yorker* not already published key sections of the text? Would *Silent Spring* still have become a best seller, absent its serialization? Perhaps. But by having shaped the magazine into an influential machine of taste for a postwar liberal class, White prepared an audience for one of the most important environmental texts of the twentieth century. She did so by normalizing concern for human health and environmental quality in gardens, backyards, and suburbs and packaging such environmentalist concerns in a digestible form: literary cultivation delivered via a national magazine.

In subtler ways, White encouraged the *New Yorker*'s readers to accept environmental critiques of chemicals. As Bailey had done in his essays questioning whether chemical sprays really represented "progress in horticulture," White anticipated Carson's concerns about the wide use of poisons in gardens. Her gardener's eye led outward to recognize that wildflowers were also threatened by herbicides that were then in use along state highways. In June 1962, a week before serial publication of *Silent Spring* began, White reviewed the year's garden catalog and declared: "We have never used chemical weed killers" and "I agree with most of the White Flower instructions [for maintaining a lawn], taking strong exception only to its advice on weed killers" (166, 167). Her critique of using chemicals was embedded in aesthetic values that are at once traditionalist and linked to class and gender, as when she parodied the catalogs selling power equipment to "suburbanites," primarily men, noting: "Having made this investment [in a riding lawn-mower] he'll almost surely want such attachable extras as a fertilizer spreader ($24.75) . . . and a Magic Fog, Jr., for broadcasting poisonous weed killers and insecticides (a mere $11.95)" (171). So when White proposed, in jest, that goldenrod be named the "national flower" for its ubiquity and endurance, she also noted the ubiquity of an environmental threat. At the end of a geographical tour, she slips in an argument for a chemical-free continent and a rejection of public herbicide spraying: "Descend into a bog and there, growing wild, is goldenrod; climb a mountain and there, between the crevices of boulders, is goldenrod; follow the shore of the sea and goldenrod gleams along the edge of the sands; drive along our highways from coast to coast in August and September and the fields and ditches are bright with goldenrod, unless the state you are driving through has destroyed them with chemical sprays" (185). The critical edge in White's style represents a new constellation of literary and environmental values, one opposed to blunt political or instrumental uses of writing, preferring instead literary indirection, parable, and a refined tone—precisely the tools Rachel Carson deployed in *Silent Spring*.

For White, literary cultivation was pragmatic and liberating in at least one other important way. As a college-educated woman, she faced pressure at the start of her career as editor in 1925 to put her own work aside to care for her children and otherwise support her husband's work. In her first year at the *New Yorker*, White published an article entitled "Home and Office," which appeared in a special issue of the magazine *Survey Graphic* dedicated to the "question" of women in the workplace. In it, White advocates against views that sought to stifle women seeking leadership roles at work; she describes her current work as "a literary profession of some sort" and emphasizes that it complemented—indeed, enabled—her role of being a better wife and mother (318). Thus the category "literary" also functioned within a gendered division of labor—literary work offered empowerment with the compromise of accepting an appropriately gendered role. White understood the "literary profession" in empowering terms: it gave her greater autonomy and avoided more limiting gender roles. An apt comparison can be drawn between White and Carson, who first pursued graduate work in biology before finding it financially necessary to discontinue doctoral research to take care of family. Carson became an editor at the Fish and Wildlife Service and, through a career as a popular science writer, developed both literary influence and an environmental critique.

This strategy or adaptation of a literary formation as environmental and political work also required innovation beyond established forms of writing, as witnessed in White's sympathetic attention to marginal writing in garden catalogs, manuals, reference books, and illustrated guides. Still, White's generous view of what garden writing might become in America went only so far. The politics of garden writing in the early postwar era, as evinced by White's reviews of seed catalogs, flower arranging, botanical illustrators, and Elizabethan garden writers, seem reticent and even conservative when placed beside contemporary geopolitics: decolonization struggles, the wars in Korea, Vietnam, and elsewhere, and repression of the political left (including the Nearings) at home. The traditionalist, liberal politics and characteristic literary style of the *New Yorker* (bemused, ironic, even-mannered) became an essential part of the dominant postwar literary formation alongside mainstream, aesthetic environmentalism. To put it otherwise: more Americans came to subscribe to the *New Yorker* edition of gardens and environmentalism than to the radical political ecology of the Nearings.

The aesthetics of White's garden writing reflect boundaries of culture, politics, and environmental thought as reproduced by dominant literary institutions. Such boundaries were productive and not merely constraining. A structuring structure, the mechanisms of taste fostered and activated a strong environmentalist critique of chemicals, as in *Silent Spring*. While the early postwar decades saw changing relations between American social classes and between these classes and the land, the dominant institutions that condition literary taste—schools, universities, and to an extent, publishing houses—rejected a more crudely materialist awareness, expressed by writers like Borsodi and the Nearings, of how food gardens might sustain healthier lives and more just communities. The numbers of magazine circulations offer a simple comparison of the relative audience for radical ideas about organic gardening for subsistence versus the audience for aesthetic ideas about pleasant and secure ornamental lawns: readership grew slowly for *Organic Gardening* from 1945 to 1980, whereas it quadrupled for *House and Garden* and expanded across the country for the *New Yorker*. There, Katharine White's garden essays found a national readership.

Looking forward and outward from White's writing and cultural context, a garden vision preoccupied with flowering borders of forsythia and goldenrod embroidery dedicated to Tennyson has held limited appeal for twenty-first-century societies facing environmental crises. Certainly an affluent town-and-country lifestyle has become practicable for fewer and fewer Americans, to say nothing of the vast majority of the planet living on far more modest ecological and economic resources. In the decades of deindustrialization and declining real wages between 1970 and 2010, cultivating gardens for subsistence and a counterculture of garden manifesto writing have spread. Political and discursive struggles over environmental justice became coextensive with hundreds of community gardens across American cities. Urban community gardeners, as discussed in chapters 5 and 6, also called for revolutionary gardening, although their activism rarely seemed directly influenced by the Nearings. A clear line can be traced, however, from Katharine White's literary cultivation and environmental critique to Michael Pollan's *Second Nature: A Gardener's Education,* to which I now turn.

3 BEING THERE, SECOND NATURE, AND THE GARDENER AS PRAGMATIST

The garden writers considered so far distinguished the ethical and political significance of gardening in terms of concerns central to American environmental thought, though not ones traditionally associated with the discourse of wilderness. Instead, they focused on healthy food, domestic economy, fair access to useful land, everyday beauty, and economic justice for small agricultural producers. For Helen and Scott Nearing, gardens served as the keystone of a life that resisted the militarism and consumerism of dominant American culture, whereas for Katharine White, gardens were a form of cultivation parallel to good literature. Their garden writing represents a less often recognized but widely influential cultural form of environmentalism and a forerunner for ideas of environmental equity and justice. This chapter turns to the ways in which pragmatic political and ethical discussions of gardening, particularly in Michael Pollan's 1991 book *Second Nature: A Gardener's Education,* have included several dimensions of environmental justice while neglecting the limits of a single gardener's perspective for recognizing difference.

Pollan's *Second Nature* presents an early version of a garden-centered ethic in what became in the 1990s a chorus of criticisms leveled against American wilderness environmentalism. At the time, the book offered a popular equivalent to the academic interventions of the "great wilderness debate," which redrew the theoretical boundaries of environmental politics, history, and literary criticism. This debate involved Ramachandra Guha, William Cronon, and other scholars who exposed the ways in which

American intellectuals and environmentalists, in their research practice, activism, and language, long imagined the natural and social worlds as starkly divided into uninhabited wilderness and the industrial conurbations of modernity. Pollan expresses a skepticism of wilderness-focused environmentalism, but on a pragmatic political basis rather than primarily in terms of a material historical or postcolonial account, as was the case with Cronon's essay "The Trouble with Wilderness" (1996) or Guha's "Radical American Environmentalism and Wilderness Preservation: A Third World Critique" (1989).

Environmental politics in the United States was far from moribund at the time of the debate. Academic and institutionalized politics simply had not kept pace with the proliferation of grassroots activism in the 1970s and 1980s that led to the emergence of the terms *environmental racism* and *environmental justice* in the work of the sociologist Robert Bullard. Environmental justice concerns entered literary and cultural criticism a decade after they gained currency among sociologists and political scientists. The first pulse of book-length treatments in American literature and culture, such as Joni Adamson's important reconstruction of a genealogy of environmental justice in American Indian literature, appeared after 2000.

The relatively recent timeframe of cultural studies focusing on environmental justice is significant for considering the literature of gardening that appeared at the end of the twentieth century. *Second Nature*—sometimes mistaken as being published closer to Pollan's best-selling critiques of the food system, *The Omnivore's Dilemma* (2006), *Food Rules* (2009), and *Cooked* (2014)—situated itself as a liberal call for moderation in environmental politics when it first appeared in 1991. Pollan's cultivation narrative ought to be seen as an attempt to circumvent the environmental backlash of the 1980s, the attack on environmental protection led by advocates of laissez-faire economics. That political backlash had succeeded in large part through a campaign depicting environmentalists as radical preservationists who cared more about endangered owls than working people. In *Second Nature*, Pollan pursues a pragmatic middle way between these two imagined extremes, primarily by appealing to a new ethos rather than calling for collective political action. The result is ambiguous. The book develops abstract ideals in universal terms through its cultivation narrative, often with blind spots to real power differences. In "The Idea of a Garden" chapter, Pollan asks directly, "What if now, instead of to the wilderness, we were to look to the garden for the makings of a new ethic?" Such a garden ethic "would not necessarily supplant the earlier one, but might give us something useful to say in those cases when it is silent or unhelpful" (225). He then presents his "provisional notes" on the "fresh metaphors about nature" needed to reframe public discussion of environmental problems as a series of ten principles.

In the two final and most comprehensive principles Pollan declares that nature "is above all a pragmatist, and so is the successful gardener," and he argues that culture, as much as nature, ought to inform an environmental ethic (231). In part because of the role that pragmatism plays rhetorically and conceptually in the book, and in part because of its liberal, ameliorist politics, *Second Nature* has become of wider interest to philosophers and legal theorists as well as to gardeners, critics, and environmentalists. The book's success parallels the ascendance of pragmatism among political liberals at the end of the Cold War. The appeal of its garden ethic succeeds at the expense of greater attention to social differences, wherein pragmatic ethos and private character trump environmental justice as a political agenda necessarily attune to inequity. My reading is consonant with arguments made by critics in food studies, including Julie Guthman (2011), Alison Alkon and Julian Agyeman (2011) and Alison Carruth (2013), about similar blind spots in Pollan's later best-selling books, although I contend that *Second Nature* marks an important renewal of writing about gardens. By contextualizing Pollan's first book-length nonfiction account as a product of its political moment—the triumphant early years of U.S. hegemony after the fall of the Berlin Wall in 1989—and its intellectual culture—represented by Richard Rorty's brand of pragmatism—I hope to make clear how his ethical vision of gardening reaches toward, but falls short of, articulating fuller demands for environmental justice.

Late twentieth-century American garden writing remained a synthetic, hybrid form in its mixture of personal cultivation narrative with broader political and cultural claims. Pollan joined the Nearings and Katharine White in offering a down-to-earth philosophy with a large grain of ironic humor and critique of American culture and environmental politics. The appeal of his garden ethic is inseparable from the literary structure of his book—and taken in the arc of this study, it can also be seen as sharing qualities with the ethical claims made by earlier and later garden writers. Pollan offers ethical insights by historicizing garden cultivation and garden writing within an intergenerational family narrative, which functions ideologically in public debates over conservation and land use. His later contributions to debates over food, agriculture, and environmental issues more generally stem from the cultural, historical dimension of environmental thought in the reinvigorated liberal, pragmatic terms of *Second Nature*. Environmental historians and historical ecologists had been engaged in similar work within academic contexts for two decades before Pollan's memoir was published; Pollan crystallized the significance of historicizing nature as entangled with human culture at the end of his narrative with a striking description of "a garden as a kind of blooming archive, a multicultural, transhistorical crossroad" (265). The image is analogous to the call Alison Alkon and Julian Agyeman have since made in their book *Cultivating Food Justice* for a "polyculture," even as they take Pollan's later writing to task for its inattention to a "privileged positionality" (3) or the limited (though universalizing) perspective he wields as a white male intellectual in U.S. society.

The problem of positionality and perspective is more complicated still. Pollan frames his "archival" view of gardening within a reflexive cultivation narrative, and a fuller evaluation of the garden-centered ethics he proposes and the political stakes for such an ethics ought to attend to this reflexive, narrative context. Such a contextual reading reveals ambiguities and ambivalences of social class, race, and gender that also affect ongoing conversations about environmental ethics and politics. These include ongoing debates over the relative importance of identity versus class politics within ecocriticism and the environmental humanities. Analogous debates over economics, health, and cultural identity occur in food studies, a field that produces its richest work through a reflexive, process-oriented, culturally embedded view of food. Pollan developed a similarly dynamic and reflexive view of gardens in *Second Nature*. His autobiographical account of making a garden on a former farm in Connecticut foregrounds how class and cultivation become linked. The text also responds to contemporary demands for social justice in a globalizing multicultural society by constructing the "gardener" as simultaneously nonideological, consensual, and neutral as well as materially engaged with—and therefore a worthy arbiter of—environmental debates.

Second Nature is an exemplary *liberal* cultivation narrative in several ways. First, its layered, ironic authorial voice belies the simplicity of its primary thesis, that a "garden-centered ethic" will have more to say than wilderness-centered ethics about how to use nature less destructively and to act conscientiously toward the natural world. Certainly, this idea was not new for leaders in conservation organizations, professional ecological restorationists, and conservation biologists, but the critical success of *Second Nature* broadened public discussion by calling readers to shed unenlightened prejudices. Further, by using comedy as a rhetorical strategy, Pollan also models ways in which scholars as well as journalists could combine ethical and political criticism with environmental history. The political significance of *Second Nature* becomes clearer through comparison with literary antecedents, such as Henry Mitchell's "Earthman" garden column and Jerzy Kosinski's novella and film adaptation *Being There*. In addition, *Second Nature* exemplifies the ambiguous relationship of white liberals and the emerging environmental justice movement in the brief moment "between the wars" (1989–2001), the Cold War and the War on Terror, a rhetorical context of a growing neoconservative movement and so-called third-way liberals led by then-president Bill Clinton. This moment witnessed a resurgent interest in cultivation of character as well as moderate environmental protection—both were cultural, reformist responses to more radical political demands.

Environmental justice activists gained national attention in 1991 with the First National People of Color Environmental Leadership Summit in Washington, DC, which joined disparate communities and voiced common grievances within a countrywide framing of justice and rights of citizens. Participants pressed for an ambitious agenda and compensation for minority communities victimized by toxic landfills, uranium mining, an asthma epidemic linked to air pollution, and unequal learning outcomes for poor African American children poisoned by lead-laced soils. More important, they called on mainstream environmental organizations and traditional supporters of civil rights, including the Democratic Party, to turn their attention to ending environmental racism. The 1992 election of President Clinton depended on support from the constituents of this national environmental justice movement. His administration seemed responsive. Clinton's Executive Order 12898 of February 1994 established the National Environmental Justice Advisory Committee (NEJAC), an advisory board with no enforcement powers. Activists have since argued that NEJAC effectively coopted the environmental justice movement (Schlosberg 202). Thus a mainstream (white) liberal agenda offered formal recognition of existing environmental racism as a strategy for achieving wider support without fundamentally changing how decisions (such as the siting of waste facilities) would be made.

Such political dealings are often described cynically as "pragmatic," but in his 1989 book *Contingency, Irony, and Solidarity,* the philosopher Richard Rorty formally divided public liberal values (e.g., equal access to amenities and fair exposure to risk) from private ironic awareness of the contingency of justice on flawed individuals and existing structures of inequality. Pollan's persona and critical vocabulary in *Second Nature* clearly echo Rorty's brand of pragmatism. Moreover, the limits and possibilities of his proposed garden ethic match the split Rorty advocated: Pollan's gardener is a public liberal and private ironist. Particularly, Pollan's garden vision reproduced Rorty's emphasis on the contingency of language and his conceptual division of the self into private skeptic (irony) and public liberal (solidarity). In sum, *Second Nature* exemplifies both an environmental pragmatism and liberal politics of gardening by advancing two claims: (1) gardening broadens awareness of the contingency of our use of the natural world, heightening our private awareness of the ironies of cultivation; and (2) gardens and the role of the gardener appeal to people across backgrounds, building solidarity or a broader consensus for politics than does preservationism. More generally, *Second Nature* taught readers to take greater interest in new modes of cultivation and be more imaginative in adapting their cultural images of nature—often by assuming a consensual relationship with readers that skipped over deeper rifts of class, race, and gender visible in the literature of gardens written by less privileged authors.

Turning to Pollan's *Second Nature* after reflecting on ideological divides over small-scale postwar gardening in the texts of the Nearings and White throws into relief how narratives of self-cultivation continue to assume (rather than foreground or even demand) the relative economic autonomy of the gardener and writer. In particular, the ways in which Pollan characterizes himself in class terms even while historicizing the cultivation of his gardens exposes a structural contradiction in American political culture. Garden writing as liberal cultivation narrative emphasizes individual sovereignty and private property but leaves its readers with an obligation to navigate plural versions of cultivation among different classes. Some of these may be deceptive pastoral fantasies; others are democratic utopias worth struggling toward.

Foregrounding a closer reading of Pollan's text with Jerzy Kosinski's deliberate political satire of garden rhetoric will bring into focus whether and how, as the legal and political theorist Eric Freyfogle has argued in a critique of Pollan's book, "tend-the-garden reasoning overlaps with today's pro-business, libertarian calls for privatization and for unleashing the free market" (1002). A comparative analysis of the gardener *as* a character in *Second Nature* and Kosinski's political fiction *Being There* is useful for theorizing how cultivation relates to cultural capital and political power in ways that open up rather than foreclose questions of culture in relation to environmental politics. Kosinski's *Being There* presents a naive American gardener as a voicebox for corporate greenwashing in Washington, DC, in the last decade of the Cold War. For Kosinski, the thin metaphorical language of cultivation is ideological cover for the maintenance of financiers' political power. In contrast, the thick biographical narration of cultivation in Pollan's book, published a year after the end of the Soviet Union, presents gardening as a less dogmatic and worldly approach to environmental problems.

Cultivating the Gardener

In the 1979 film adaptation of the novella *Being There*, Peter Sellers plays a middle-aged gardener named Chance who, after his employer's death, leaves for the first time the private walled garden that he has maintained his entire life. A wealthy socialite's chauffeur promptly backs up a limousine and pins the bewildered Chance against another parked car. Because Chance is dressed in a tailored suit left to him by his deceased patron, he is mistaken for a financier who will likely sue over his injuries. The socialite cannily invites Chance to be examined by her husband's personal physician at their home. Her husband, Benjamin Rand, is an ailing businessman and presidential advisor. A series of comic misapprehensions unspools when

Rand mishears and renames Chance the gardener "Chauncey Gardiner." Thus elevated, Chauncey's banal observations about gardening are taken as an ethical and political vision, which eventually gains the ear of the president. Garden imagery and metaphors link Chance, Rand, and the president as these characters exchange and generate political power through apparently everyday discussions of watering, pruning, and plant growth. Kosinski's original text and its screen adaptation lampooned this language of cultivation as a basis of power rooted in the shared tastes of a ruling financial elite in late twentieth-century America.

In the book more so than in the film, Kosinski suspends the question of whether Chance/Chauncey is an admirable, childlike sage or merely a simpleton granted proximity to America's financial oligarchs. The suspense builds as Chauncey's platitudes about gardening enter the print and televised mass media cycle: through a casual reference in the president's economic policy speech; through an interview on a late-night talk show; in the editorial pages of the national papers. Comments such as "As long as the roots are not severed, all will be well . . . in the garden" seem to promise hope (and fiscal growth) in the declining economic context of 1970s stagflation. Speaking of gardens offers hope to all without promising anything to anyone.

In addition to ridiculing the potential emptiness of garden metaphors in late twentieth-century American culture, Kosinski satirizes the tendency of democratic discourse to yield to naturalizing abstractions when it seeks "common ground." Using such language, he takes aim at a central contradiction of American political culture, presented in a memorable aside that Rand delivers to Chance: "Your position in the financial community carries a lot of weight. But what draws [the president] to you is your down to earth philosophy." Again and again in the country's history, liberal democratic aspirations to "down-to-earth" equality have contested with a capitalist political economy led by financial oligarchs. In Kosinski's ironic creation, the character Rand imagines Chance's financial connections—they share purely speculative wealth—and thereby manufactures the gardener's political influence. The critique of Chance's "down to earth philosophy" as a garden ideology thus provides both a formal and a historical perspective on the development of American garden writing after the 1970s. As we have seen, that "progressive" tradition of writing from

Liberty Hyde Bailey to the Nearings to Katharine White placed gardens at the center of environmental awareness and social justice. But how clearly could a garden-centered critique of widening economic inequality and environmental degradation be received in a media climate impatient with anything longer than an earthy-sounding prescription for growth?

Kosinski reveals how ahistorical garden metaphors continued to appeal to democratic sentiments while supporting financial elites. The political satire par excellence of the language of gardening, *Being There* is also a critique of authoritarianisms of both socialist and capitalist varieties during the Cold War. The film visually connects the history of garden design and political ideology in the final scene of Benjamin Rand's funeral, set at the real-life Biltmore Estate in Asheville, North Carolina. Beginning in 1888 Frederick Law Olmsted designed the estate for the family heir George Washington Vanderbilt, including a mausoleum modeled on the burial place of the eighteenth-century German garden designer Prince Hermann von Pückler-Muskau. Although the princely estate provides a visual grandeur to impress upon viewers the fictional Rand's power, the setting also alludes to Olmsted's complicated legacy as an elite designer of public landscapes. One recent biographer, Justin Martin, has pointed out that Olmsted pursued the commission for the Gilded Age millionaire's estate because he anticipated that the resulting landscape would influence Vanderbilt's powerful visitors. "This is to be a private work of very rare public interest in many ways," wrote Olmsted of the Biltmore project (qtd. in Martin 363). In a manner more blatant than Central Park, whose history of dispossession preceding the park's creation has been reconstructed by the social historians Roy Rosenzweig and Elizabeth Blackmar (77), Biltmore promoted top-down landscape design as a way to harmonize social differences and, hence, naturalize a political order. Such a garden landscape in the background of Rand's funeral serves as a stage set for democratic culture, a Progressive Era design occupied by new financial oligarchs and their party apparatus. In the foreground of the scene, Rand's pallbearers discuss how to engineer Chance's selection as the next presidential candidate to ensure continuity of their political influence.

In addition to highlighting this ideological function of garden language and design, *Being There* focuses the political work and conservative cultural connotations of gardeners, and in particular the way in which "the gardener," singular and abstracted, often serves class interests. As we have seen, American garden writers from Bernard M'Mahon onward sought to shape the character of their readers through practical advice, assertions of taste, and humor, often even in nationalist terms. The most widely published American garden writer contemporary with Kosinski's novel produced a philosophical forerunner for Chance. Beginning in 1970, Henry Mitchell wrote a Sunday garden column for the *Washington Post*, which he signed "Earthman," and published his collected essays in 1981 as *The Essential Earthman*, with later collections in 1992 and (posthumously) 1998. In hundreds of columns in the country's second-largest daily, Mitchell dispensed advice and philosophical reflections, representing himself as an ironic yet powerfully normative figure.

Though Mitchell frequently addressed the reader from a confident first-person perspective, he also referred to himself at times in the third person, as "the gardener." "The gardener" was by turns bemused, cantankerous, and capable of obsessive absorption in judging cultivars of azaleas and roses. In style and temperament, Mitchell's gardener reiterated many of the comic conventions used by the long-time *House and Garden* editor Richardson Wright and the Czech novelist Karl Čapek, whose *Gardener's Year* was translated into English and published in 1931.

Unlike his predecessors in the mold of Liberty Hyde Bailey's "progressive gardener," Mitchell's gardener more often than not sought light-hearted humor, quiet pleasure, and gratifying beauty rather than broad democratic lessons, agrarian justice, or scientific horticulture when he wrote of his home garden in Washington, DC. A Southerner and something of a reactionary in temperament, Mitchell ridiculed urban neighbors who had what he called "the grubby-peasant approach to life" and planted fruit trees with "the thought of saving a dime a year by eating wormy produce instead of buying it from good farmers" (195). Mitchell's "Earthman," like Chance the gardener, presents an abstracted, archetypal gardener at the edge of a postindustrial city—an individualist rather than communal figure, standing at odds with the 1960s and 1970s counterculture of gardens and food activism that Warren Belasco investigated in his 1993 study *Appetite for Change*. Mitchell's figure of the gardener as somehow both cultivated yet elemental—an earthman rather than a caveman—has proved of perennial interest. The gardener becomes a figure of consensus by appealing particularly to white middle-class Americans' unspoken investments—what George Lipsitz called the "possessive investment in whiteness"—the structuring postwar fictions of universal affluence based on property and democratic culture.

Pollan's *Second Nature* resonated as a narrative of cultivation within an established cultural context in which the gardener served potentially as a pragmatic crank or political naïf for greening up the status quo. Given Pollan's role as a leading voice of the alternative food movement and the influence of his later best-selling books on public discussions of food justice, a more contextual reading of his self-presentation is long overdue and the stakes quite high. For example, Pollan's open letter in 2008 to the U.S. president-elect presented an uncanny parallel to Kosinski's fictional scenario in which a gardener gained the American president's ear; among other suggestions, he called for planting a garden again on the White House lawn. First Lady Michelle Obama's establishment of such a garden in 2009 appeared to materialize the values of the new administration by tapping into the consensual political symbolism of gardening. The garden's stated purpose was to serve as a progressive model for addressing a public health crisis in childhood obesity, which disproportionately affects the country's poor and minority communities. The historian Cecilia Gowdy-Wygant has since observed that a gardening first lady appealed to "attitudes about women's responsibilities to feed and care for children's health" (181) and evoked nostalgia for victory gardens during the deepening economic crisis of 2009. The food studies scholar Julie Guthman has offered a more trenchant analysis of the political theater of the White House Garden, linking it to Pollan's attempts to move the countercultural critique of the food system into the mainstream. "An approach that appeals to all parts of the political spectrum," writes Guthman in *Weighing In: Obesity, Food Justice, and the Limits of Capitalism,* "cannot challenge the political-economic forces that are producing cheap, toxic, and junky food" (186). Similarly, conceiving of gardens and gardeners as a material and symbolic strategy for addressing health disparities, environmental injustices around urban decay, and more fundamental social inequities presses against limits of political economy.

Making characters by literary cultivation is political work, whether we are inclined to see literature as potentially liberating or ideologically constraining. Pollan's garden writing offers a model articulation of environmental ethics but requires critical reading of how it coordinates private selves and public duties, private property and public environmental quality. One way to understand more critically the political dimension of Pollan's language is to compare the way the gardener is constructed as a character in *Being There*, in which the American corporate oligarchy relies on public misreading of private gardening as blankly ideological, perpetuating a dysfunctional alienation of citizens in an age of television. By contrast, Pollan's comic, quirky narration of himself as gardener imagined new directions in politics and ethics to avoid the dead-ends of wilderness environmentalism.

Though rhetorical analysis allows for comparisons of character narration across genres, fiction and nonfiction work differently to cultivate the reader's character by modeling political attitudes through patterns of action, including situational irony. Our most cherished views of the good life and our smallest gestures are formed partly by reflecting on the styles of life and the processes of reflective minds (or the absence of reflection) represented in both literary types. Critical differences exist between Kosinski's gardener and that of Pollan, differences made possible both by the charged context in which each author wrote and by the literary forms in which they worked. *Being There* calls readers to interpret gardening language in political discourse cautiously because its very figurative richness makes it super-serviceable for authoritarians.

The novella that furnished the later film is more thematically structured around television and gardening as mediations of the postindustrial economic slowdown. Chance is obsessed with watching television and learning how to shape his appearance and behavior according to others' expectations; at every point in the novel, his listeners compulsively attribute figurative meaning to his literal descriptions of gardening. These literal gardens belong categorically to formal, semiprivate outdoor spaces manicured by professional staff, yet they range from the Old Man's garden, where Chance had worked his entire life before the opening of the novel's action, to the luxurious though indistinct garden outside the United Nations reception, which figures at the end of the book. Figurative gardens come to refer to the state of the global economy, specific phases of human life, including death, the vagaries of the publishing industry (104), and, finally, the biosphere endangered by industrial toxins (106). Misreadings of garden language rip away its environmental referentiality, specific gardens yielding to a universal discourse of power and position.

Chance's character functions as a screen, compared with the absorptive surface and horizontality of TV from the first scenes, upon which others project their desires, beliefs, and ideologies. Before leaving the Old Man's garden and entering a picaresque televised journey, his existence is undocumented and he is illiterate. Other characters consistently mistake his illiteracy for reluctance, or in the case of KGB and CIA agents investigating his identity, for intrepid secrecy about his identity. The KGB nicknames him "blank page." The book also alludes to the seriousness of the broader historical context in the early 1970s to underline the trivial nature of Chance's garden talk and mediated image: an American economy in "stagnation" and "decline" as well as a continuing war in Vietnam. "To millions, the war, I suppose, is just another TV program," a young woman tells Chance at a party, "but out there, at the front, real men are giving their lives" (107).

The political satire of *Being There* depends on two types of political ignorance. First, Kosinski represents the U.S. political economy during the Cold War as dominated by a know-nothing capitalist oligarchy. The backroom governance of Rand and his associates is ironically homologous to Soviet rule by party elites, represented in the novel by the Soviet ambassador and KGB agents. Second, American citizens—excepting the novel's readers, presumably—are poor interpreters of political rhetoric, mistaking figurative garden-related language for economic prediction and preferring appealing physique to more problematic characters. The book assumes that critical reflection, of the sort involved in reading serious novels, might lead to more skepticism about the political status quo.

The political satire of *Being There* works insofar as readers know not to read Chance's garden talk as political speech, understand enough to distinguish between a language literally descriptive of caring for private gardens and a figurative language of cultivation that, because Chance has no conceptual reference except television for launching his metaphors, takes on whatever meanings his audience desires. As readers, we are invited to enjoy the situational irony of powerful businessmen who mistake a mentally impaired gardener for a financial seer. Nowhere is the irony more didactic than in the scene with Rand and Chance following their meeting with the president: "Chance, you don't play games with words, you're direct.... Your position in the financial community carries a lot of weight. But what draws Gary [the president] to you is your down to earth philosophy" (screenplay). Chance's garden "philosophy" consists of descriptions that are so general and ahistorical, they lend themselves to the "games with words" that the financial community and, in turn, the president use to present themselves as "down to earth," connected with their broader audiences.

In the rough economic times alluded to in the literary fiction, audiences are most often represented as seeking pastoral comfort. Chance's language, with its flat descriptions of gardening work (e.g., "The garden needs a lot of care"), becomes political speech when its auditors misread it as a fulfillment of pastoral wishes (55). The president opportunistically repeats the limited stock of garden platitudes that Chance says during their meeting. By the novel's end, the oligarchs are enthusiastically proposing Chauncey Gardiner as a potential vice-presidential candidate; in the 1979 film, it is implied that Chance will become the presidential candidate. *Being There* warns against the danger of gardening entering political discourse as a source of dead metaphors, as figures of speech that reinforce the status quo—in other words, the danger of cultivation as ideology. The financial community stands behind Chance's down-to-earth philosophy precisely because of the duplicity of cultivation as a class-based agenda enacted through earthy language.

Although gardening and a language of cultivation evoke class distinctions, significant differences separate the language of cultivation as it relates to character in *Second Nature* and *Being There,* two philosophically inclined, popular literary creations. In the latter, class connections, physical attractiveness, and fluency with television empower the political speech of the main character. The film adds white privilege to the list. Chance's former caretaker, a black woman named Louise who knows he is cognitively impaired, sees him on television and declares, "All you got to be is white in America to get what you want." In contrast to the film's candid moment of critique, in the novella Kosinski only once invites readers to identify with the perspective of a character other than Chance. Having assumed the public persona Chauncey Gardiner, confidante of financiers and the U.S. president, Chance appears on a popular nightly talk show. When he questions the meaning of the host's inquiries about the economic recession, the host guides him toward elaborating on the metaphor of the garden used by the president in his most recent speech on the economy: "'I know the garden very well,' said Chance firmly. 'I have worked in it all of my life. It's a good garden and a healthy one. Its trees are healthy and so are its shrubs and flowers, as long as they are trimmed and watered in the right seasons. . . . And there is still plenty of room in it for new trees and new flowers of all kinds'" (55).

Some in the audience clap; a few others boo loudly. The host turns toward the camera and says: "It is your view, then, that the slowing of the economy, the downtrend in the stock market, the increase in unemployment . . . you believe that all of this is just another phase, another season, so to speak, in the growth of a garden . . ." ([ellipses in original] 56). Kosinski presents the televised interview from third-person indirect discourse inside the unreflective mind of Chance, then repeats the scene from the first-person limited perspective of a lawyer, Thomas Franklin. Franklin is watching the televised interview at home with his wife. Echoing the program's host, Franklin's wife interprets Chance to be predicting a recovery in the economy and she criticizes her husband's forlorn mood about their finances. The scene ends with the closest articulation of a fixed norm for the satire from Franklin's world-weary perspective: "'Like a garden.' He sighed audibly. Sure. If one could only believe that" (59). Up to this point, the appeal of Chance's garden philosophy has depended on a suspension of disbelief shared by his (mostly white) audiences—perhaps Chance is a sort of sainted idiot gifted with prophecy? But Franklin's outright rejection deflates the garden metaphor because he hears the naturalizing language as wishful thinking in the context of more sobering economic anxieties.

The gardener/narrator of *Second Nature* dispenses with assumptions that gardens are natural and invites scrutiny along with intimacy. Pollan organized the book as an imaginary tour that travels from his wealthy grandfather's posh gardens to his father's Long Island suburban lawn to his own five-acre garden on a restored farm in Connecticut. In this account, gardens tell personal histories, stories about flawed people bargaining with the natural world and one another; these stories occasion refinements of guiding metaphors. In *Being There*, gardens are essentially reduced to spaces of capital, from the walled urban garden of an American businessman to Rand's palatial gardens, the American imitations of European aristocratic landscape architecture. *Being There* takes as its fictional garden aesthetic a stylized, European picturesque and uses the impaired consciousness of the curatorial Chance to sensitize readers to the deceptiveness of garden tropes. *Second Nature,* by contrast, narrates a "garden tour" of upwardly mobile and unabashedly bourgeois yards in its story of the cultivation of the author's character. Pollan's cultivation narrative pointedly dispenses with anodynes about gardening in hard times.

The broader context of world politics also changed from the time *Being There* and *Second Nature* were published, such that garden-centered language and ethics came to have rhetorical weight in American environmentalism; Kosinski's parody corroded faith in a language of cultivation aimed at altering politics. The political satire of *Being There* culminates in the final scene, in which the financial oligarchs discuss putting Chance on the presidential ticket. Its thrust is that the thin metaphor of America as a "garden"—we might think of the pastoral imagery of Ronald Reagan's "It's Morning in America" advertisement, released the same year—deceives because it means something to everyone but offers no challenge to status quo corporate oligarchy. The satire works because we are invited, by means of omniscient narration disclosing Chance's cognitive limitations, to ridicule those who read Chance's language as metaphorical. The film's iconic final sequence shows Chance walking across an ornamental pond at Biltmore Estate, a beguiling image of pastoral democratic leadership and commentary on the fantastical element in U.S. presidential politics.

The differences between Pollan's gardener persona and Kosinski's Chauncey Gardiner point to a substantial gap between a garden ethic involving reflective self-cultivation and a pejorative version of cultivation as pretense and fantasy. Chance is a suit without a reflective mind, sense of place, or historical consciousness, a true "blank page." Pollan's narrator is careful to emphasize the historicity of each garden he has made: the palimpsest of his Connecticut garden with the cultural models of his family's gardens and the farmer's use of the land before him. Chance is also primarily a television persona, whereas Pollan presents a character tightly woven into a reflective literary tradition of American garden writing. The success of Chance's garden aphorisms, which read flatly on the page, is due to his physical attractiveness and mimicry of television characters. Even the most prescriptive moments in *Second Nature* place Pollan the gardener and writer in a literary genealogy that is canonical and even Anglo-American, as when he describes himself as a "child of Thoreau" and cites Pope's *Epistle to Lord Burlington:* "Consult always the genius of the place." Where Chance confirms whatever people thought before hearing him—his tropes map onto the economic optimism of the president, the pop spiritualism of his studio audience at the talk show, the literary pretensions of the Russian ambassador—Pollan's gardener invites readers along in a process of education that involves both environmental historical knowledge and elite cultural literacy.

Beginning with the earlier history of the Connecticut farm, Pollan the gardener upends both the popular fantasies of heading back to a pristine nature and the notion that a language of nature exists outside history and politics. He further demonstrates this point by applying hyperbolic analogies drawn from political history to his education as a gardener. This education seems to be as much about how to become an eccentric liberal-minded ironist as it is about gardening. A second-order irony arises when Pollan represents learning to garden as synonymous with learning to abandon received ideas about nature and the human place in it—while relying heavily on a Europhilic tradition of ideas of gardening that hardly reflects the diverse historical trajectories of gardens in the United States.

Environmental Pragmatism, Ethics, and Liberal Politics

At its best, garden writing bridges ethics and politics, engaging social lives, natural systems, and political economy. What is it about gardens that challenges assumptions of private selves and private spaces and renders ethical questions more coherent and their political stakes more glaring? Perhaps it is their liminal character, the way even private gardens always border on public space, crossing the two categories. *Second Nature* presents the author, an educated gardener, as a comic hero and ethical guide capable of reconciling contradictory properties of American landscapes. Toward the book's end, Pollan shifts from writing in the first person to describing an ideal type in the third person, "the gardener." To a large extent, "gardener" is synonymous with "environmental pragmatist" or even "cultivated liberal ironist." Drawing this connection more explicitly between his rhetoric and Richard Rorty's terms in *Contingency, Irony, and Solidarity* helps clarify the conceptual strengths and a few weaknesses in the garden ethics of *Second Nature* as way to engage environmental politics. Particularly, Pollan's narrative persona, eerily similar to the implied author of Rorty's book, also manages the split personality of a critical liberal subject, aware of the disunity of garden and gardener, private property and public character. In American culture, self and property have been understood in the Lockean liberal political tradition and popular imagination as mutually constitutive. Autobiographical garden writing is a form well-suited to reproducing the interpenetration of selfhood and property.

Amanda Anderson's discussion of literary self-cultivation as a mode of reflective reason helps clarify the most important difference between Pollan's and Kosinski's gardener. The capacity of reflective reason is precisely what Chance lacks and what Pollan's gardener epitomizes. The gardener's capacity for critical reflection, in other words, is central to the relevance of *Second Nature* as an articulation of environmental ethics. Pollan's character is a pragmatic citizen of a democratic polis, ready to participate in a procedural democracy for managing nature; Chance embodies a desire to appear well on television while naturalizing whatever business wants for his popular audiences. Anderson's assessment of pragmatist "characterology" illuminates other rhetorical similarities between Pollan and Richard Rorty, including the narrative structure and disruptive ironic histories in *Second Nature*. Evoking a model character and moral psychology is a signature rhetorical strategy that Rorty borrows from William James when he describes the ideal type for the pragmatist toward the end of *Contingency, Irony, and Solidarity*. Pollan turns most explicitly to both the vocabulary and projection of character in Rorty's book. Whereas Pollan would have us believe that "Nature is a pragmatist," it is more instructive to read this line as a projection, wherein pragmatic attitudes in the gardener are assigned to "Nature," who indeed resembles the character of Rorty's fictional pragmatist: adaptive, sanguine, as happy to deal in beauty as well as truths, and always recognizing the contingency of cultivation and clime.

Before analyzing the particular structure of Pollan's cultivation narrative, it is useful to dig more deeply into the connection that Amanda Anderson has made between pragmatism, character, and public debate. Her chapter "Pragmatism and Character" in *The Way We Argue Now* provides a suggestive framework for thinking about the way the persona of Rorty's "pragmatist" is spoken through the implied author of *Second Nature*. Anderson reminds us that William James, in his lectures on pragmatism, describes the "dispositions" of philosophers: the "tender-minded" and "tough-minded" rationalists and empiricists. He then proposes pragmatism as not only a middle path and "mode of thought" but also, in Anderson's view, as "a full-fledged personality, a feminized mediator and reconcilor." Rather than presenting an abstract theory of pragmatic thought or a methodology, James models a way of thinking, an attitude of radical openness within a hypothetical pragmatist thinker, presented in feminine third-person, and then, according to Anderson, he "brings his heroine to life through near-literary description" (115). She faults Rorty's approach in dividing public and private for how "it narrowly reads character out of the liberal principle of tolerance, and it consequently forecloses the characterological dimensions of liberalism's dedication to individual flourishing, ongoing critique, and openness to difference (rather than mere toleration of it)" (133). Anderson would have us "distinguish between pragmatism's psychology ... explicitly avowed and theorized, and its characterology, which remains gestural, descriptive, and implicit, typically not fully integrated into the logic of the argument" in order to better flesh out the significance of character for debates over ethics (116).

Anderson's distinction of this so-called characterology of pragmatism also illuminates why Pollan places such weight on describing the character of the gardener toward the end of "the Idea of a Garden." In his description of the gardener, he reverts to referring to "Nature" with a feminine pronoun. This pronoun use and style is uncannily similar to William James's lecture as well as Rorty's character sketch of the "liberal ironist" versus the "liberal metaphysician" in *Contingency, Irony, and Solidarity*. For Rorty, the metaphysician emphasizes truth in abstract propositions and grand theories to defend liberal political principles, while the liberal ironist thinks "the good liberal has a certain kind of know-how," and her particular skill is "recognizing and describing the different sorts of little things around which individuals or communities center their fantasies and their lives" (93). This passage marks a clear conceptual trajectory from James's earlier characterization of Nature as a pragmatist through Rorty's liberal ironist to Pollan's emphasis on characterizing a set of attitudes and interrogating ideas of wilderness, nature, and the garden that are central to American liberalism.

The neopragmatists' "gestural, descriptive, and implicit" notion of the liberal ironist informs the narrator's purpose in *Second Nature*, in particular his interest in gardens as places replete with individual and collective meaning. Pollan's focus on interpreting gardens in their everydayness resembles what Rorty characterized as the liberal ironist's strategy of "recognizing and describing the different sorts of little things around which individuals or communities center their fantasies and their lives" (93). Pollan's ideal gardener is presented explicitly as imitating Nature, a personified abstraction who in turn is described as a pragmatist, thus forming a self-enclosed conceptual loop. Furthermore, the book pursues a strategy of changing the vocabulary in which we think about the environment, providing fresh metaphors, embracing contingency, and savoring details of different gardens. Indeed, Pollan's whole narrative of cultivation pairs an ironic awareness of his family's heritage as gardeners and developers with liberal hopes for reinventing environmental politics.

Pollan applies Rorty's suggestion that, facing a conceptual impasse, we shift from "one metaphoric" to another. Rather than reassigning "truth-value" so that everyone will now agree that, actually, Nature is a garden and Nature includes people, Pollan attempts at the level of autobiographical garden writing to "get from one vocabulary [e.g., of wilderness and industry] to another [e.g., gardening], from one dominant metaphoric to another" (Rorty 48). Pollan also displays an impatience with "metaphysics" and foundationalism, rejecting attitudes associated with deep ecology implicitly by advocating for "an ethics of use" and frank "anthropocentrism" (227). Pollan demonstrates a pragmatic commitment to literature as a way of thinking that can change people's lives by giving them a new vocabulary for seeing the world, not insight into the foundations of things "as they really are, out there" in the world. Finally, he converts his private history as a gardener into a form of bildungsroman that created this persona, this "gardener" conscious of the environment and human uses of it in contemporary culture.

Recognizing where the explicit garden ethics of *Second Nature* is bound up with the neopragmatist's assumed characterology returns us to the analysis of character in literature *as* ethical criticism, a task that from Wayne Booth's *Ethics of Fiction* through Amanda Anderson's *The Way We Argue Now* has been reported to have "suffered a kind of exile from theoretical work in the field of literary and cultural studies" (Anderson 134). One of the ways to further our critique of character in *Second Nature* is to concede that an author's voice achieves moral authority through literary strategies that are adaptive to a specific political context and more or less attuned to reach a particular reading audience. What is the relationship between the ethical claims in *Second Nature,* the conciliatory, welcoming, ironic, and self-deprecating narrator, and the historical context in which these claims developed? What are the limits of this pragmatic retreat into liberal character? Are these limits symmetrical with the affluent white-male proprietary perspective of Pollan's gardener?

Already in the 1990 *Harper's* forum discussed in the introduction, Pollan had begun to voice skepticism about whether wilderness environmentalism had not already reached its imaginative and political limits. Yet the very name of that forum—"Only Man's Presence Can Save Nature"—reveals the extent to which its participants had not absorbed the feminist critique of gender and environmental domination made by pioneering critics such as Carolyn Merchant and Donna Haraway in the 1980s. Moreover, the forum participants (all white, all male, all over thirty years old), while worldly and progressive-minded, nonetheless represented a narrow swath of perceptions and historical experiences of nature. Similarly, although Pollan's perception and guiding aesthetics are tinged with an ironic sensibility in *Second Nature,* garden cultivation is finally a figure of self-cultivation, an activity linked in the United States with property ownership and white privilege.[1] Therefore, and perhaps despite his liberal intentions, a second-order irony of unexamined privilege is revealed when Pollan writes that "the margins of our gardens can be tropes too, but figures of irony rather than transcendence—antidotes, in fact, to our hubris. It may be in the margins of our gardens that we can discover fresh ways to bring our aesthetics and our ethics about the land into some meaningful alignment" (300). This garden and gardener nexus, with its qualities of irony, assumptions of property and identity, and tension between rough and refined land, is fertile ground indeed for figuring psychic investments in environmental issues, or the engagement of character and landscape.

However, the limits of privileged perspective do not render Pollan's work irrelevant to environmental justice. *Second Nature* is about becoming a gardener as an appealing form of "self-creation," in the spirit of Rorty's liberal ironist, and sharing a public-spirited role—becoming a gardener as a way to act responsibly as a citizen. Pollan exemplifies a trend to change the meaning of *garden* and *gardening* in American culture so that the constellation of terms familiar to us from environmental politics, words like *toxicity, pollution, wilderness, preservation,* and *conservation,* extend their range. One risk in reading Pollan is to overstate the public influence his work might have. In contrast to Rorty's halving of the contemporary psyche into private ironist and public liberal, Pollan's narrator describes gradually "aligning" the real and ideal gardens.

Not only are irony and contingency shaping concepts throughout *Second Nature,* but the book also demonstrates a faith in literature and creativity to forge solidarity, the conclusion reached by Rorty in *Contingency, Irony, and Solidarity* that has since been developed in forms such as Martha Nussbaum's revival of liberal humanism in *Poetic Justice* (1995) and subsequent work. Pollan's *Second Nature* models thoughtful, reflexive engagement with culture, history, and environment by linking self-cultivation and garden cultivation, enriching the traditions of American garden writing inherited from Katharine White and the Nearings. I have argued that Pollan's garden-centered ethics presents a conventional, if attractive, character in the gardener and writer, cast in the philosophical mold of neopragmatism, especially the emphasis in Rorty's pragmatism on the contingency of language. Both can now be understood as the contingent products of a brief period of American hegemony between the Cold War and the War on Terror. We should be cautious about the historical limitations and possibilities of this characterology, particular as it models the relationship between ethics and politics.

Shifting attention to the role people play as designers and nurturers, as potential gardeners who make "a legitimate quarrel" with nature and seek fruitful accommodation rather than outright domination (Pollan 229), may not alter the equation of power, property, and environmental use in America. An environmental politics informed by pragmatic, character-based ethics depends on existing liberal institutions and beliefs; it does not found them. In this regard, Eric Freyfogle is right to warn readers of *Second Nature* that viewing the world from the scale of one's private garden might only reiterate "the American liberal ideal of live and let live—precisely the attitude that has brought such destruction to our lands" (1001). He is right primarily because such an ethic is bound by the limitations of unequal social processes of inscribing selves. Liberal ideals of selfhood in America have been closely aligned with owning land, and environmental and conservation policy in America has been furthered by collective action. The political vision of *Second Nature* is liberal in its ameliorist and reformist emphasis on character and moderate change. It advocates less destructive ways of using nature through private cultivation and spreading this ethos like a treasured heirloom, rather than calling for radical structural change in land, agriculture, economics, and environmental policy. The value and limitations of *Second Nature* are thus circumscribed as much by its structure as a cultivation narrative as by the institutions of liberal politics in America.

Pollan's "education of a gardener" publicizes these private stories of gardening, infusing remembered gardens with rich details and bringing their historical role to the foreground. In *A Philosophy of Gardens,* his 2006 book on gardening, ethics, and the good life, the British philosopher David Cooper begins by disputing a tendency in Western philosophy to exclude gardens from ethical consideration. Along with "what falls outside the 'on-duty' sphere of life," gardens have been lumped with other private spaces: "Providing no rights are violated, no moral principles broken, then what people do in their dining-room, bedroom, or garden is nobody else's moral business: ethical considerations do not apply" (88). It is by bringing to public relevance "the [private] garden's contribution to what the first of those authors [Pliny] called 'a good life and a genuine one'" that Cooper imagines that the significance of gardening for ethics may be articulated. It is precisely the public value of private garden experience that Pollan enacts in *Second Nature* by narrating his garden as a public intervention in environmental discourse.

Writing in 1991, Pollan presented gardening as constructive for American environmental politics because public debate of the environment in the 1980s had been framed in oppositional terms in mainstream media, in which hypothetical "radical environmentalists" were supposedly facing off against a countermovement of ranchers, developers, and the lobbying organizations of extractive industries. He also proposed a garden ethic at a moment of dramatic geopolitical change: *Second Nature* appeared at the close of the Cold War and the beginning of increased awareness about the consequences of anthropogenic climate change. James Hansen, a NASA climate scientist, had testified before Congress during the heat wave of 1988 about the threat of global warming, and Bill McKibben, then a young staff writer for the *New Yorker* magazine, had already declared the "end of nature" (1989). *Harper's* 1990 forum, from which Pollan developed his book, devoted most of its pages to covering the collapse of Soviet satellite states in Eastern Europe. An ethic centered on the idea of a garden and informed by gardening practice presented a liberal, cosmopolitan ethic, arguably one that turned a critical gaze inward to the process of cultivating American identities.

In a more globalized context for environmental politics, a garden ethic seemed more consistent and more practicable than one rooted in deep ecology, which, as the environmental historian and public intellectual Ramachandra Guha warned, tends to re-create ecological imperialism and third world "frontiers" preserved according to first world conservation prerogatives (75). In contrast to deep ecology, a garden ethic would be more logically consistent by systematically including people as part of nature and avoiding the schizoid landscape of market and wilderness. And it would be more practicable, since those of us indirectly pillaging the world live far from the dwindling ark of "wilderness," the 8 percent and shrinking portion of untrammeled land. Insofar as the book rose to meet its aim of shifting environmental ethics away from deep ecology and wilderness, *Second Nature* was also a political project, a sharing with readers of the imaginative and practical work of garden cultivation. The book exemplifies a writer's pursuit of better modes of cultivation: accounts of how genre and practices of writing engage with and are shaped by environmental contexts, both social and nonhuman, and a pragmatic perspectivism that endorses humbler forms of anthropocentrism.

Cultivation Narrative and Weakened Anthropocentrism

If Pollan's garden-centered ethic feels timely and compelling two decades later, it is because his eloquent narrative of self-cultivation scrutinized the evolution of environmental values of the postwar American middle class and exposed persistent contradictions in its environmental politics. Rather than presuming his readers to be virtuous environmentalists eager to put nature ahead of human interests, Pollan insinuated the cultural value of particular, remembered gardens as a template for a more generous and restrained sense of human agency. He did so in pragmatic and humorous prose, shifting discussion away from theoretical distinctions toward deliberation of the practical consequences of human uses of the natural world. Pollan's critique in *Second Nature* was also effective within the bounds of mainstream discourse because, like his later popular successes *The Botany of Desire* and *The Omnivore's Dilemma*, it did not press its critique of American consumers too hard. As a professional journalist and public intellectual, Pollan has spoken especially to the rhetorical dynamics at work in deepening an environmental ethos in American culture. In this sense, too, a garden-centered ethic offers a compromised intervention, a way to address the epistemological problem of human perspective in ecological knowledge and the ethical problems of responsibility to biophysical and social others.

One of the primary values of Pollan's cultivation narrative is how it instantiates a responsive, hopeful, but also more chastened human perspective toward nature as a gift. David Cooper has unpacked the ways that Pollan's description of his garden as a gift relates to its ethical value. The logic of the gift as unexpected, unpaid-for benefit underwrites Cooper's claim that making a garden induces certain virtues. Gardening requires developing greater attentiveness, according to Cooper, so it can be said to give or grant several virtues to the paradigmatic gardener. Cooper uses Pollan's description of growing an heirloom Sibley squash as an extended illustration of four virtues that he argues are "induced" by gardening: care and respect for life, structure and routine, humility, and hope. If the gardener returns each year to plant, believing the seeds will grow, then she already exhibits what Cooper calls the "pre-condition virtue" of hope: "Unless confidence is invested in the power of virtuous practices to be conducive to the good life, there could seem no point in them. Hope is not simply induced by this or that garden-practice, like growing squash, but it pervades the very ethos of gardening" (96). Although it is easy to dismiss hope as a vague or sentimental factor in any environmental ethics or politics, increasingly scholars have turned to recognizing the importance of our emotional responses to whether we address serious environmental problems, deny, or avoid them.[2]

Cooper represents the total effect of the virtues induced by gardening as an "unselfing," a phrase he borrows from the novelist Iris Murdoch that echoes Kantian "disinterestedness." In environmental ethics, "unselfing" more readily maps onto nonanthropocentrism, biocentrism, and ecocentrism. Here, Cooper's ethics of gardening contradicts Pollan's more modest claim that a garden-centered ethic would be "frankly anthropocentric" (227). At the same time, Pollan argues that gardening invites a widening of human interests to include nonhuman life, a more enlightened and broader sense of human interests. A more succinct description of this ethical position is what the ethicist Brian Norton has called "weak anthropocentrism" (163). As another philosopher, Andrew Light, describes it, "weak anthropocentrism" describes two facts: we make our ethical judgments from an unavoidably human perspective; yet by imagining the interests of nonhuman parties, we diminish or "weaken" the limiting effects of this anthropocentrism.

In contrast to Cooper's abstract reading, Pollan provides a model for a gardener in the world aware of environmental history as second nature rather than *Urwelt* or ideal. This approach does not detract from Cooper's general claim that gardening is pertinent and even central for a good life, broadly conceived to include the costs that life exerts on an ecosystem. However, Pollan's text presents a historical awareness of complexity in his own gardening; even the history of the Sibley squash enriches its meaning for the narrator.[3] When read in the fuller context of the narrative, we recognize that Pollan exalts the squash as a way to resist the dominance of reducing all values to market values. This is not only a generic feature of capitalism but also a psychic inheritance from his grandfather's entrepreneurial gardening.

Pollan describes his grandfather's garden as a place where vegetables became markers of exchange value and a means for him to maintain his paternal influence well into old age. As "an eighty-five-year-old real estate magnate," he would harvest his vegetables and go "on his rounds—visiting tenants, haggling with bankers and brokers, buying low and selling high—he'd give away baskets of vegetables" (20). These are not gifts in the purified sense he later ascribed to the Sibley squash: "But the remarkable fact is, my Sibley, considered from the vantage of the entire planet's economy of matter, represents a net gain. It is, in other words, a gift" (172). The gift in this restricted, asocial sense is framed as an inversion of the second law of thermodynamics: "Plants . . . are energy returned to matter—entropy undone, at least here on Earth" (173).

Through the unscientific trope, Pollan undermines market economics with a moral economy. Compared to this gift, by which "the second law of thermodynamics is repealed" in his garden, which Pollan calls "the harvest's most salutary teaching," his grandfather's gifts appear secondary and tainted. For as Pollan had learned in his education as a gardener, "grandfather never gave away anything that didn't have at least some slender string attached to it." Rather, his grandfather's gifts of produce to business partners in the real estate market, those "sweet-as-sugar beefsteaks," are strategies that he imagined "put these businessmen in his debt, gave him some slight edge," a way to "put the suits off their guard, making Grandpa seem more like some benign Old World bumpkin than the shark he really was" (21). Pollan's "salutary lesson" from growing a Sibley squash is richer in hope and pleasure—two undervalued elements of any environmental ethic—than his grandfather's narrow cash nexus. This lesson results from a materialist history of gardening in his family rather than an idealist ethics centered in gardening.

The ethic is embedded in an American cultivation narrative. From the first page, Pollan establishes this dual process of cultivation—both changing the land, "five acres of rocky, intractable hillside in the town of Cornwall, Connecticut" and changing himself: "This book is the story of *my* education in the garden" (1). Moreover, the goal of the cultivation narrative is to bring the ideal garden and the actual place into closer "alignment." The narrator establishes himself on intimate terms by describing his long personal history with gardening and a now-familiar trajectory of American intellectuals who left Manhattan for the New England countryside. This shift from metropolis to "a sliver of a derelict dairy farm on the eastern edge of the Housatonic Valley" is described as a conscious decision that creates the conditions of possibility for gardening: an advancing forest confronts the newly exurban landowner with a necessity. The forest "did in fact threaten to engulf our house," so "I had to do *something*—either mow the weed patch that passed for a lawn, or put in a real garden" (2). The author is motivated by "the recollected satisfactions of childhood gardens" to attempt the process of cultivation. More specifically, Pollan describes having made "the wager . . . that gardening might be worth taking seriously, and that, closely attended to, it might yield some good stories and helpful ideas" (6).

To garden, then, is to enter history via the narratives implied in the dual process of cultivation, or as the landscape architect Anne Whiston Spirn has eloquently described it, to learn to speak and write the "language of landscape." For Pollan's narrator, getting into the garden also means confronting an attachment to stories that do not fit his context. That context is both a personal and an environmental history, moving from a Long Island suburban childhood through the metropolis and into a postagrarian New England landscape. The overarching story is one of how a character, by creating a multistaged garden landscape and comedy of cultivation, follows "the logic of [his] experiences," which leads him to the "idea of a garden—as a place, both real and metaphorical, where nature and culture can be wedded in a way that can benefit both—[which] may be as useful to us today as the idea of wilderness has been in the past" (6). He then structures the narrative as a patrilineal trajectory of establishing his own private garden that in turn serves as an ambivalent marker of masculine maturity and landed property.

Critical cultivation requires scrutiny of intergenerational family narratives. Pollan meditates on two gardens in his family's history: his father's suburban lawn and his grandfather's grander spread in Babylon. This approach invites intimacy and a psychological interpretation of his motives as a gardener: the two properties have financial histories that connect, albeit tensely through debt obligations, his "indoor dad" (11) and his paternal grandfather, a Russian immigrant who, "starting with nothing, selling vegetables from a horse wagon . . . eventually built a fortune, first in the produce business and later in real estate" (12). He interprets his grandfather's garden as a sign of Old World virtues rendered visible in statuary, borders, and productive vegetable plots. His father's garden also seems to represent a kind of success, or the imperatives of hard work and a defiance of the convention of postwar lawns as a sign of national belonging and conformism.

From the Long Island front yard that Pollan's father neglects, partly in defiance of his fastidious neighbors, the family moves up to a fancier, estate-like house in a newer suburban development. There his father hires professionals to maintain the landscaping. These early anecdotes prepare us for a narrative that repeats in some form this patrilineal pattern of upward mobility. They also underline the psychological dimensions of real estate, the way his grandfather's cultivation enables acquisition of property and depends upon it. The narrator implies a tension in his grandfather's practices and views: "Some days my grandfather made the earth speak vegetables; other days it was shopping centers." When he writes that "in his mind, the Old World peasant and the real estate developer existed side by side; he was both men and perceived no contradiction" (13), we understand the potential contradiction is between a developer's view of land as a source of speculative wealth and a peasant's attachment to a particular patch of earth. Pollan's narrative of cultivating his own five-acre garden tracks his ambivalence to his grandfather's dualism as gardener and real estate developer. But it is also a response to his father's rebellion and, later, his compliance with the American lawn as a cipher for political conformity.

Chapter 1 of *Second Nature*, "Two Gardens," begins the cultivation narrative by opening a space between ideas and ideology, or ideas and their political uses, and this space is at once conceptual and genealogical. For "Two Gardens" reads like the secret history of private gardens, the details of a sacred place, an "enclosed and privileged space out-of-doors," to echo Pollan's first definition of a garden (7). The story involves romance and antagonism with paternal authorities, and its intergenerational masculine tensions shape the character that Pollan cultivates, the *ethos* he presents of himself as a gardener.

His grandfather represents ambivalent ideas of "loving the land as a source of value," whether as a farmer who sees possible cash crops or as a real estate developer who would buy and resell land for suburban homes at a profit. The rivalry between Pollan's grandfather and his father is presented in part in class terms: his father refuses to behave as a "renter" and his grandfather attempts to assert himself as a "landlord" by lending to his son the down payment on the Long Island house. Ownership of the garden is an axis of tension early in the story. In Pollan's genealogy, outside is "Grandpa's realm ... where he and his gardener, Andy, had made what I judged a paradise" (16). Pollan describes this garden in lush detail, leading to a passage in which he reads the underlying family romance at work in his recollection:

> The area between the lawn and the beach was twenty or thirty feet deep, thickly planted, and it formed a kind of wilderness we could explore out of sight of the adults on the patio. Here were mature rhododendrons and fruit trees, including a famous peach that Grandpa was said to have planted from seed. It was an impressive tree, too, weighed down in late summer with bushels of fruit. The tree was a dwarf, so we could reach the downy yellow globes ourselves. Hoping to repeat Grandpa's achievement, we carefully buried the pit of every peach we ate. (Probably it was his example that inspired my experiment with watermelon seeds.) But ripe fruit was only one of the surprises of Grandpa's wild garden. There was another we always looked for, only sometimes found. Creeping among the rhododendrons and dwarf trees, we would on lucky days come upon a small, shaded glade where, on a low mound, a concrete statue stood. It was a boy with his hand on his penis, peeing. This scandalous little scene never failed to set off peals of laughter when we were in a group; alone, the feelings were more complicated. In one way or another, Eros operates in every garden; here was where he held sway in Grandpa's. (17)

Placing the nude statue at the end of the long, graphic passage begs the question of where else Eros has been promiscuously transcribed in Pollan's depiction of the garden, as in his boyhood memory of the peach's "downy yellow globes." Eros is at the heart of the energies in Grandpa's garden in other images of control and satisfaction, such that "the [vegetable] garden looked like nothing so much as a scale model for one of the latest suburban developments: the rows were roads, and each freestanding vegetable plant was a single-family home" (18). His grandfather's desire to be treated as a landlord structures the linear grid of his garden and his habit of calculating the value of the vegetables that he gives away. Like other gifts, they come "not unencumbered," but enable "feelings . . . more than a little proprietary" toward the recipients. Paternalism describes this combination of male gifting and proprietary feeling.

By bringing such a fatherly relation to light as a factor in environmental choices, Pollan's cultivation narrative seeks to preserve the transmission of his grandfather's gardening knowledge without unwittingly reproducing his more confining gender and class roles. Paternal, entrepreneurial capitalism characterizes his grandfather's attitude toward nature explicitly: "To work in his garden was to commune with nature without ever leaving the marketplace" (20). Being in Grandpa's vegetable garden, surrounded by the "wild garden" with its "sacred grove" to Eros, sets the stage for Pollan's suggestion that "the idea of a garden" might resolve the conceptual contradictions between laissez-faire economics and wilderness preservationism in symbolic terms. The intergenerational self-awareness that Pollan generates through the cultivation narrative is a prerequisite for his resulting garden ethic.

From ChemLawns to a "Blooming Archive"

Second Nature demonstrates an ironic awareness of how the aesthetics and ethics of gardening reinforce familiar gendered narratives. The gap between appearances and reality in Pollan's tour of his own garden is most poignant in how it retraces a masculine rivalry between his father and grandfather. He links his own ambivalence to suburbia and the conformity of chemically fed lawns to his father's refusal to mow and maintain a front yard at his childhood home on Long Island. When pressed by his neighbors to conform, his father angrily storms out and mows his initials into the knee-high grass. This "angry grass graffiti" admitted "a whole other set of meanings, having to do with social or even political questions." The "unmowed front lawn," Pollan writes, "was a clear message to our neighbors and his father-in-law" (23). He reads his father's avoidance of lawn care as ambivalence toward "the suburban way of life" presenting a false sense of classless democratic equality in America. The reflection opens into a general meditation on suburbs. He calls them "part of a collective landscape" (24), neither public nor entirely private, and cites Lewis Mumford's 1938 description of suburbs as "a collective effort to live a private life" (*The Culture of Cities* 215). (It is worth noting that Mumford was referring specifically to the romantic suburbs of the late nineteenth century.)

Pollan describes his father as "eager to sign up" for the suburb's linkage of family and private property, yet in the laborious maintenance of front lawn conformity, Pollan sees that the "whole middle-class utopian package" also required psychological "dues to pay" (24). Thus he ties his own narrative of cultivation as a revision to the psychological costs of maintaining the civilized appearance of a middle-class society in America. He wants to let go of the "vaunting or brutish sign" associated with his father's resistance to conformity, but he is equally wary of his grandfather's sensibility, which he imagines would reject as unprofitable the "bog . . . on the other side of the fence" on his farm in Connecticut (299). The tensions are eventually displaced onto his garden, where nature returns with a feminine pronoun:

> Here is *my* autograph, then, scratched on this green page [of meadow] with my roaring Toro, this very prow of culture! This stylus of Western will! Yes, the grasses have their days, too, coming usually after a drenching rain, when the ink on the signature blurs and the fresh green growth dissolves my path's keen, whetted edge. No more than nature herself, these grasses haven't the least regard for any human form or identity, for even the greenest of our thoughts. But so what?! I can remow my path every week. . . . Nature does tend toward entropy and dissolution, yes, yes, but I can't help thinking she contains some countervailing tendency, too, some bent toward forms of ever-increasing complexity. Toward us and our creations, I mean. Toward me and this mower and the otherwise unexplainable beauty of a path in a garden. (304)

Pollan leaves us with a series of images that undercut masculine mastery, an agrarian fantasy of the plow of culture mastering nature.[4] He adds an ironic turn to these tropes, such that the phallic connotations of his roaring Toro dissolve in the palpable self-mockery of hyperbole: "This very prow of culture! This stylus of Western will!" Yet rather than taming an imagined wilderness with his plow, he is keeping the "conversation" open, acknowledging that "nature is bound to have the last word—will be forcing fresh shoots out of the ground long after I've ceased to weed or mow." Pollan repeats the formula in the final pages, describing the gardener's relation to nature as "like the argument of old friends, or husbands and wives after long years, a quarrel renewed . . . with no end in sight, and no end sought" (303). Such language intimates that Pollan thinks of making gardens not as a sign of male mastery of the world or a status symbol, but rather as the work of Eros: a contacting of forces and energies driving toward greater complexity of forms and animation.

Such disputes set the stage for *Second Nature* and come full circle in the book's final chapter, "The Garden Tour," with Pollan mowing a path in the restored meadow of his Connecticut garden. Both the meadow and the mown path are a complicated generational rescripting of his father's derelict Long Island lawn. The tour is also a kind of return, a narrative form well suited to changing our conceptual—and perhaps practical—relationship with the natural world. If making gardens is paradigmatic of an environmental ethic, by creating a "place that admits of both nature and human habitation," such work will require not "a harmonious compromise between the two" but "continual human intervention" (59). An important dimension of finding the proper character of this intervention is the psychological working through of inherited values, particularly the coproduction of gender roles, class, and historical landscapes that Pollan undertakes.

Cultivation, as both gardening and literary practices, is never far from class in *Second Nature*. Although the narrator's environmental values cannot be reduced to a result of his own privileged biography, the rise of more collective forms of gardening in the 1980s seems in retrospect conspicuously absent from the book's wide-ranging interests. This broadening and deepening of ecological interests among gardeners and environmental justice activists is surprisingly not registered as part of Pollan's critique of wilderness. The charge of elitism has frequently been leveled against environmentalists since the 1980s, in particular from environmental justice activists. Yet it has most often, ironically, been launched from those on high in corporate offices, as Philip Shabecoff has observed in his history of the environmental movement, *A Fierce Green Fire* (2003). Cultural elites certainly played a key role in the preservationist environmental movement of the early twentieth century, but by its close, environmental politics had broadened into grassroots activism as well as mainstream interest-group strategies. As with environmentalism, so with gardening in America: the American Community Gardening Association grew from forty-nine founding members in 1979 to over half a million by 1999 (Dawson 238).

Pollan balances an aesthetic critique of the false democracy of the suburban lawn by demonstrating affinity for this middle landscape. At times his tone contains more than a little nostalgia for his affluent suburban childhood—and ultimately *Second Nature* asks the reader to put lawns in their place by reducing their size and recategorizing home ground as a give-and-take with nature rather than a "symptom of, and a metaphor for, our skewed relationship to the land" (76). His remedy is not less or more cultivation but different creative uses of nature and culture: "If lawn mowing feels like copying the same sentence over and over, gardening is like writing out new ones, an infinitely variable process of inventions and discovery" (64). Appealing to individualism, he writes that creating a garden via enclosure, a kind of "secession from the national lawn," can make "a distinct and private place" (65). Pollan's Connecticut garden reconciles the contending ideas of lawns and gardens, suggesting that we renegotiate the class politics of cultivation home by home. Pollan repeatedly evokes the way lawns materialize American myths of classlessness, such that a continental greensward signifies a democratic leveling among middle-class suburbanites.

What about those of us without the means or space to enfold a rose garden into the design of our home grounds? That is one area where Pollan's proposed garden ethic requires elaboration, a reckoning with how landlessness figures into garden praxis, as we will see in the making of community gardens in chapter 5. The stories of community gardening supplement rather than substitute for Pollan's work in *Second Nature* because his garden ethic is above all pragmatic and pluralist. It is in the final garden tour that the hybrid design of his Connecticut garden, which contains a perennial border, rose garden, herbal *hortus conclusus,* rows of heirloom vegetables, and a mown meadow of native grasses, makes visible an inclusive "ethic" for using and caring for the environment, one that imagines environmentalism as horizontal, pluralist politics. Perhaps Aldo Leopold was right that we can be ethical only toward land that we see, feel, understand, and love—and private ownership may not be a precondition for caring use. In the postindustrial, globalized present, we experience land in divided patches with distinct interests, all mediated by culture.

Thus we might hope that more Americans can learn to cultivate pluralist narratives about land and invest in more creative options than the orderly, herbicide-dependent verdant chemical lawn. Or, as Pollan suggests, that we might see our work as the making of "a kind of blooming archive, a multicultural, transhistorical crossroad" (265). Such garden cultivation is not only metaphorical. Rather, in the climax of *Second Nature,* Pollan materializes the plural community in horticultural terms: "Sibley squashes once cultivated by American Indians root in the same soil as the Madame Hardy rose, the white damask bred in 1832 by Monsieur Hardy, gardener at Malmaison. Welcome here, too, in this ordinary patch of Connecticut soil, will be the honeyed, green-fleshed Jenny Lind, beloved muskmelon of the nineteenth century, cast aside by commerce. . . . And here, too, rising like conning towers over my garden will be hollyhocks I've grown from seeds that I traded my Sibley's for, and which my correspondent tells me were harvested from plants whose seed she obtained from a painter on Long Island who gathered *his* seeds from Monet's garden at Giverny" (265). By embedding this mental image of his garden as a "blooming archive" and as "a teeming, polyglot free port city where all manner of diverse and sundry characters" meet, Pollan defines the role of the gardener as part librarian, part pollinator of new ideas and plants (268). This call to creative participation is itself a significant political intervention and turn in environmental ethics.

As we will see next, Pollan's is not the first utopian, democratic vision of gardens in America as sites of intercultural exchange. For Southerners such as Elizabeth Lawrence and Alice Walker, garden cultivation was an ambivalent site for confronting racialized divisions of labor as well as an activity that connected women across class and generational lines. Yet the public-spirited, pluralist, weakened anthropocentric ethic has, through the charismatic persona of Pollan's more recent criticism in *The Botany of Desire*, *The Omnivore's Dilemma*, and *In Defense of Food*, begun to leave its signature on the political landscape.

4 RACE, REGIONALISM, AND THE EMERGENCE OF ENVIRONMENTAL JUSTICE IN SOUTHERN GARDENS

Michael Pollan's image of his Connecticut garden as a "blooming archive, a multicultural, transhistorical crossroad" (265), tapped into cherished hopes about America as a multicultural nation, which he transferred to the space of his garden and then transposed to a universal garden ethic. Pollan also presented his garden ethic as contingent and experiential, "provisional . . . [and] based on my own experiences and the experiences of other gardeners I've met or read" (225). These experiences as recounted in his book *Second Nature* occur in gardens made in relative peace and affluence. In even more precise terms, Pollan's call for a garden-centered ethic seems bound up with the environmental and literary history of New England gardens and writers from Henry David Thoreau to Katharine White.

By the late twentieth century, New England was well on its way to becoming a vast second-growth forest with fewer working farms and less industry. Moreover, a tradition of nature writing in the region, dating from Thoreau, had established a national audience and publishing market for environmental writing. The landscape of rural Maine, Vermont, and Connecticut bore traces of longstanding inhabitation well suited to confirming a view of the wider landscape as a vast garden, culminating in a garden-centered ethic that could risk importing the self-confirming worldview of a relatively affluent and homogenous class. The environmental and cultural conditions in which White, Pollan, and Scott and Helen Nearing gardened are regionally specific and historically contingent. All gardened on former small New England farms; all migrated to the country from New York, their writing enabled as much by connections in the city's publishing market as by the relatively cheap acreage in the region's hinterland following the contractions of agricultural and industrial activity. These garden writers expressed self-awareness of class and gender, but they did not face racial prejudices or more desperate struggles for land and means of cultivation. Widening the geographical reach of this study in this chapter and the next reveals significant regional variation and socioeconomic conditions at work in American garden writing. In southern garden writing in the last century, the cultivation of both garden and gardener was often racialized, however unconsciously.

Gardens cultivated in the southern states and the texts they inspired inevitably return to landscapes of plantation agriculture and tenant farming. Southern garden writing might confront or suppress structures of institutional racism that operated in the wider landscape, but it could not escape them. Southern gardens became "multicultural crossroads" with ambivalent histories. They include dooryard and subsistence plots grown by enslaved workers, what Judith Carney and Richard Rosomoff call "botanical gardens of the dispossessed" (124), as well as the formal gardens cultivated by generations of Euro-American southerners in imitation of their idea of a genteel antebellum South. This chapter focuses on three insights into the particular modes of cultivation in southern garden writing during the last century of rapid social change while evoking the longer shadows of slavery. First, I explore the tension between segregation and exclusion on the one hand and, on the other, the circulation and movement of plant materials within a porous, but unequally accessed, regional culture of cultivation. Second is an investigation of how texts capture the racist violence displaced onto the landscape. Third, I delve into the efforts—in texts and in gardens—to produce, refine, and stabilize an idea of region rooted in the land in the face of rapid changes during the South's relatively late industrialization and the civil rights and environmental justice movements.

The particular interest of such southern writers as Julia Lester Dillon, Elizabeth Lawrence, and Alice Walker lies in how they revitalized vernacular flower gardens (and, to a lesser extent, kitchen gardens) as places perpetuating memory and where white and black southerners connected.

These garden encounters occurred in a context of structural inequality, unequal access to land, and economically generated famine that racialized hunger.[1] Reading these garden writers together reveals a persistent tension over the ways in which southern cultivation relates black and white southerners as figures in a historical landscape divided unequally through the mid- to late twentieth century. It also provides a point of orientation for more public struggles over community gardens that have occurred in recent decades.

Geographical scale has been a controversial point in literary environmental criticism, with leading critics such as Ursula Heise seeking to shift the field's longstanding interest in local and bioregional literature of place (exemplified by Gary Snyder's watershed aesthetics) toward an emphasis on the interpenetration of global processes within the local (28). Yet the ensuing discussion, while productive, has tended to concentrate on the extremes—from the local to the planetary. This tendency holds a substantial risk for historical research and contemporary criticism, for regional cultural patterns continue to influence the use (and abuse) of landscape in the United States. How, for example, has literary and cultural regionalism affected the perception of environmental racism, the unequal distribution of risk and amenity between racial groups in a shared landscape? Prescriptive and fictional representations of southern gardens make visible the ambivalent meanings of cultivating landscapes and refining perceptions, most often in eruptions of interracial tensions between southerners in the space of their gardens. Notably, black and white southerners have made gardens together, and this making involved innumerable exchanges across the color line even as the meaning of that boundary has shifted and the materials exchanged—seeds, plants, knowledge—have hardly held stable meanings.

Feminist readings of southern literary history have also made clear that southern literature has often schematized the regional culture in masculine pastoral terms, praising novels of epic scope and privileging myths of the garden. Since Robert Beverley's first published history of the Virginia colony in 1705, European American male writers have celebrated the beguiling mildness of the Southlands. "A Garden is no where sooner made than there, either for Fruits, or Flowers," Beverley wrote enticingly of the Virginia climate to prospective English settlers (79). The master trope of the garden has held sway in literary tradition at the expense of attention to the varieties of narratives and southern gardens. Environmental historians, beginning with Mart Stewart's studies of the engineering of wet-culture rice plantations in the South Carolina and Georgia Lowcountry, have shed light on the significance of Africa and gardens at the margins of plantation landscapes. Dianne Glave subsequently emphasized the vitality of post-Reconstruction gardens cultivated by black sharecroppers and rural people in North Carolina. The history of urban gardens by both the black and white middle class in the New South has yet to be written.

There is a pattern in landscape history of studying southern gardens in black and white, that is, as separate traditions. Richard Westamacott has recovered vernacular traditions particular to African American southern gardeners. Westamacott catalogs the distinctive aesthetic features of these vernacular gardens, though many of these design features are not exclusive to one ethnic group. These include brushed-dirt yards and reused materials such as painted stones and used tires in eclectic yard art. Other garden elements (e.g., bottle trees) bear traces of African cultural roots, including links to West African spiritual practices. In attending to the particular creative beauty of individual gardens, such work often misses the copresence of black and white southerners in historical gardens and their textual representations. As with much vernacular garden history, documentation of unpaid labor and casual design can be scant.

White landscape architects in the twentieth century have reinvented plantation idylls and scenes of southern cultivation through stereotypes, including those about white mistresses, often reproduced for audiences outside the South, as the garden historian Greg Grant has noted. As sites of interstate tourism, these formal gardens have higher visibility than the scratched-dirt yards of former tenant farmers. Yet the southern gardens that have become icons of antebellum plantation design have interracial histories in which knowledge of gardening was not white people's exclusive property. White slave owners relied on skilled African American gardeners as much to cultivate, propagate, and manage prized botanical collections and raise vegetables as to farm cash crops such as rice, as the historians Mart Stewart, Judith Carney, and Richard Rosomoff have documented.[2] Carney and Rosomoff in particular describe gardens of escaped slaves across the Americas, focusing on Maroon enclaves in swamps and mountains in Latin America. Such gardens—planted with maize, manioc, yams, African rice, bitter melon, and groundnut—became hybrid creations of Amerindian and African ethnobotanical knowledge. Any reconstruction of racial identities to make future lives and landscapes less segregated requires that we understand the ways southern gardens have been represented as sites and sources of cultivation. The Maroon garden as symbol of the survival of African culture through botanical fusion is one site in a whole "shadow world of cultivation" (125). Southern gardens, as part of the wider botanical archive in America, resulted from transatlantic and hemispheric exchanges with brutal consequences for enslaved laborers that are often not glimpsed in decorous accounts of plantation landscape design.

Southern garden writing also presents an exceptional case for understanding cultural regionalism at work in a language and landscape. It is often a record of environmental factors conditioning plant cultivation within a region defined by its climate, though manuals to southern gardens deploy many of the conventions of racial signification that in the last century made regionalism, and southern regionalism in particular, a suspicious project of white self-justification. Rather than dismissing regional garden writing as parochial, it is worth paying closer attention to the interaction of garden design and race as bound up with a social geography of difference that operates at many scales around the world. Moreover, in understanding one of the most popular southern garden writers, Elizabeth Lawrence, in relation to her sources and contemporaries, we may recover regionalist terms of cultivation that do not share a reactionary racial agenda. Lawrence's garden writing suggests a trajectory of widening consciousness of social connections across the South, from the fairly conventional reproduction of an unspoken white, genteel southern garden in her first book (*A Southern Garden*, 1942) to a utopian vision of free exchange in her last book, *Gardening for Love* (1987). In between, she became close friends with Katharine White, with whom she kept up an affectionate correspondence until the latter's death (Wilson *Two Gardeners*). Lawrence's blind spots to racial signification in southern gardens prepare us to better appreciate the radical cultivation of antiracism in Alice Walker's political novel, *Meridian*, as well as her well-known essays on her mother's garden and the role of black women artists. Walker's ambivalence to a rooted counternarrative to racial and environmental exploitation remains critical in imagining environmental justice for damaged lands and peoples.

In her textual cultivation of southern gardens, Lawrence reveals an imagination educated beyond bigotry or parochialism but not free of sectional blindness to the way that gardens signify constraint and exclusion as well as refuge. It is revealing to compare Lawrence's version of southern gardens with the critical views of antebellum gardens in Frederick Douglass's 1845 autobiography. (*Gardening for Love* also includes a bizarre literalist reading of Harriet Beecher Stowe's "nice descriptions of gardens" in *Uncle Tom's Cabin*.) I will frame Lawrence's work in an intergenerational context of southern garden writing, from Julia Lester Dillon (landscape designer and garden writer) in the preceding generation to Alice Walker (novelist, poet, and activist) in the following generation. This genealogy yields a more complicated and nuanced picture of southern garden cultivation, particularly floriculture,[3] as embedded in both regional culture and social history and its racial significations.

The Tarred Fence and the Boxwood Hedge

The environmental historian Ted Steinberg has dubbed the upheavals of southern environmental history since Reconstruction "the extraction of the New South," emphasizing the way that region and its people became an economically dependent source of natural resources (cotton, timber, coal) for the national economy. In the wider landscape in which logging, mining, and mechanization of agriculture were accelerating through the first decades of the twentieth century, subsistence gardens physically nourished newly industrialized workers in mines and mills, subsidizing the advance of industrialization (114).

Yet dooryard flower gardens and plantation landscapes also represented ideals of beauty and stability, as seen in the case of Julia Lester Dillon's manual *The Blossom Circle of the Year for Southern Gardens,* first published in 1922. Instead of regarding southern garden writing as a marginal literary development, I place it on equal footing with other regionalist responses to the rapid environmental and social changes in the twentieth century. Although a dominant version of southern literary regionalism has been associated with masculine pastoral, gardening (and vernacular gardens in particular) for subsistence and beauty inspired nonfiction that should be included in what Elizabeth Harrison has called the "female pastoral" tradition of fiction by southern women writers, capable of finding in gardens places of memory, autonomy, and community (13). The region's economic dependence and despoiled environment certainly do register in southern fiction, for example in the vanished forests of Faulkner's *Go Down, Moses* and the logging trucks that rumble down the roads and upset Mrs. Rainey in Eudora Welty's short story "The Wanderers," which in turn inspired Elizabeth Lawrence's late garden essays. Such fictional narratives of dependency and exploitation contrast symbolically with the values that Lawrence, Welty, and Alice Walker associate with gardens.

However, a simple gendered dichotomy oversimplifies the ways in which gardens and gardening have been represented in southern nonfiction; race has been as significant as gender in their construction. Restored plantation gardens, such as those visited in magazine garden tours and depicted in Dillon's *Blossom Circle of the Year* seem to perform a compensatory function akin to the Southern Agrarians' myth of an uncomplicated antebellum past rooted in the land, a myth of loss appealing perhaps to particular white male fantasies of power and control. The rich variety of vernacular gardens fosters perspectives in Lawrence and Walker that challenge this blanched picture of southern cultivation.

To elaborate how southern garden cultivation is constituted by racial signification, it helps to examine several moments in *Blossom Circle of the Year*, the first book-length guide to planting a flower garden in the South. Dillon, a Georgia native and landscape architect trained in Boston and New York, worked first in Augusta, designing private estates for wealthy businessmen at the turn of the twentieth century before becoming the first woman to work as superintendent of parks in Sumter, South Carolina. She published *Blossom Circle of the Year* just before assuming responsibility for planning Sumter's public gardens, a legacy still visible in the South Carolina landscape. The book is dedicated to women who work as public gardeners, and the values echo the "Social Gospel" associated with the City Beautiful movement and the Progressive Era influence of Olmsted, Bailey, and national leaders of the Garden Club movement such as Louisa Yeomans King.

Dillon's wide-ranging project of cultivation combines garden work and education. Her panoramic view of southern landscapes promises sweeping beautification, with the region treated as one enormous garden. As her dedication explains: "The women of all the South are now thoroughly awakened to their responsibility and opportunity, not only in the matter of conservation, but also in that of education and of the development of the garden and landscape work of both the cities and the rural districts." Among the works of cultivation she praises are the conversion of public spaces into gardens, in "school grounds and courthouse squares, municipal parks, and railroad stations," as well as "development, along artistic lines, of the new highways that connect the states, and . . . endeavoring to develop the farms until they become estates worthy of this or any other section" (11). Hers is at once a heavily gendered and moralized vision of reform, but the insistence on beautifying an entire landscape (including working farms) also breaks down a Victorian legacy of "separate spheres," though without ceding women's role as moral guides.

Dillon's dedication and the book itself are firmly fixed in the requirements for cultivating flower gardens in southern climates, but the drive behind cultivation is to compete with the rest of the country, to have farms equal to those of "any other section." The gardenlike appearance of farms, estates, and the boundaries of highways and cities would compensate amply for the region's underlying economic dependence, a fact made plain in Augusta. Dillon helped name Augusta "the Garden City of the South," and northern industrialists' winter homes financed the cultivated appearance of the city, which became most famous in the second half of the twentieth century for its annual golf tournament. The value of the Augusta estates' cultivated appearance traded on associations with genteel white antebellum culture, a fantasy of aristocratic gentility that, ironically enough, Olmsted's own published *Travels* had lain to waste at its origins in the mid-nineteenth century.

Dillon's reform agenda for beautifying the southern landscape is haunted by the legacy of a racist division of labor in the making of antebellum plantations. Before exploring further the ambivalence of Dillon's revision of plantation garden design, let's step back in time and consider plantation gardens from the perspective of Frederick Douglass in 1845. Douglass exposed the moral ugliness of the illusion that cultivated plantation grounds made of the South an earthly Eden. In his autobiographical *Narrative,* Douglass describes a "large and finely cultivated garden" on the Lloyd plantation as a corrupted version of an earthly Eden for whites only: "This garden was probably the greatest attraction of the place. During the summer months, people came from far and near—Baltimore, Easton, and Annapolis—to see it. It abounded in fruits of almost every description, from the hardy apple of the north to the delicate orange of the south. This garden was not the least source of trouble on the plantation. Its excellent fruit was quite a temptation to the hungry swarms of boys, as well as the older slaves, belonging to the colonel, few of whom had the virtue or the vice to resist it. Scarcely a day passed, during the summer, but that some slave had to take the lash for stealing fruit. The colonel had to resort to all kinds of stratagems to keep his slaves out of the garden" (350).

Colonel Lloyd's "most successful" technique is "tarring his fence all around," then having the chief gardener (a Mr. M'Durmond) whip any slave caught with "any tar upon his person" (351). Lloyd's plantation garden materializes so profoundly a corruption of Christian paradise that Douglass elaborated the scene in his later autobiography, *My Bondage and My Freedom* (1857), damning the system behind the landscape in his praise of its "Eden-like beauty." The critic Michael Bennett has also drawn on these passages as primary examples of anti-pastoral, with the colonel's garden read as "an allegory of the plantation" in the context of a broader "devaluation of country living" (200). While Bennett focused on demonstrating the linkage of cultural progress and urbanization in African American literature, I wish to linger on the specific critique of spatial perspective and white cultivation in Douglass's work—effectively drawing out the edge of its anti-pastoral.

In both aforementioned versions of his autobiography, Douglass draws attention to the ways in which the beauty and refinement of the gardens at Colonel Lloyd's plantation reinforced the political power of the slave-owning class. Furthermore, Douglass emphasizes that what Lloyd's enslaved workers called the "Great House Farm" was "associated in their minds with greatness" (348). The prospect was of a piece with the property system, with the landscape a projection of white ownership in the mental maps of black enslaved people. Douglass presented a seductive garden like beauty surrounding the plantation house: "The intermediate space [between the entrance gate and the Great House] was a beautiful lawn, very neatly trimmed, and watched with the greatest care. It was dotted thickly over with delightful trees, shrubbery, and flowers. The road, or lane, from the gate to the great house, was richly paved with white pebbles from the beach, and, in its course, formed a complete circle around the beautiful lawn. Carriages going in and retiring from the great house, made the circuit of the lawn, and their passengers were permitted to behold a scene of almost Eden-like beauty" (67).

Here Douglass uses language familiar to his Christian abolitionist audiences to expose the ironic moral "order" of Colonel Lloyd's garden as a microcosm for the slavery system.[4] His panoramic perspective simultaneously enacts and critiques the ways that plantation garden design controlled the prospect of white visitors as passive viewers from their carriages, as "passengers [who] were permitted to behold a scene."[5]

Much changed in the seventy years between Douglass's and Dillon's representations of southern gardens. However, Dillon's manual does not altogether abandon the perspective of the white carriage passenger described in Douglass's narrative or Colonel Lloyd's fixation with defensive borders around his garden. What remained constant was the function of gardens of grand scope to display the magnificence of wealth and thus to ratify a social order; the policing of the garden's borders became deliberately less violent, though not less anxious. Nowhere was this anxiety over enclosure clearer than in Dillon's recommendations for flowering hedges as borders. She first recommends two privets, "Amoor" (also called "Amur," *Ligustrum amurense*) and California (*Ligustrum ovalifolium*), then goes on to assess various options for "strength and durability," recommending *Citrus trifoliate* because "nothing can penetrate it 'from a rabbit to an elephant,'" as well as "a defensive hedge" of the common (and invasive) buckthorn. For "the old-fashioned formal garden, such as our grand-mothers used to make, Boxwood, *Buxus sempervirens*," she considers indispensable. Dillon remarks that in "old gardens" from Boston to New Orleans, boxwood outlined the formal beds: "In their quaint and stilted way they stand as monuments to that ante-bellum period of the geometric design and the formal garden. They belong to the day of brick paths and tangled shrubs with an Arborvitae boundary hedge, with the lower Boxwood borders outlining the designs. These old Boxwood borders are certainly attractive, the old evergreens are many of them stately and beautiful at this time, and both seem everlasting in their slow growth, but who would make such a garden now? Let us preserve these that we have, in honor of a day long dead, but for the new ones the new order to which we have changed is certainly best" (43). The boxwood hedge thus becomes not merely a physical barrier but also a symbolic prop of an old order. In suggesting both preservation and renovation, Dillon here seems innocent of the ways that landscape designs evoke memories of structures of exclusion and white power in the South.

The rather elliptical discussion of boxwood in old southern gardens does not prepare readers for the photograph on the following page titled "Old Subjects and New in an Old-Time Garden." The term *old-time* was the praise adjective of choice for the most successful American garden writers of Dillon's generation, such as Helena Rutherfurd Ely, Louise Beebe Wilder, and Alice Morse Earle. In Dillon's text, it seems to refer to the architecture and landscape design seen in the image: a classical portico, its columns partially obscured by a spreading live oak, fills the background. On the left of the walk that is central in the frame, pyramidal trimmed boxwoods line an approach to the front porch. Two young girls, one white and one black, stand in the right foreground. Dillon includes a descriptive subtitle: "Note the pyramidal Boxwoods edging the entrance walk. In the foreground are Frances with her fiery tresses and Adeline of the dusky curls—known to all the family as 'Rouge et Noir.'"

The girls' faces seem to register the kind of frustration you might expect from children interrupted in their play to become subjects of a formal photograph showing this ostensibly "old-time garden." Are they the "new subjects" referred to in the title? Is the "old subject" the boxwood, presumably a planting of some years trimmed low, or the house? Is it the overall design of the plantation landscape, with its formal approach? Or is the "new subject" the unspoken topic of race and racial relations set against a southern landscape appealing to white nostalgia?

The photograph joins the class issues, complex histories, and racial ambivalence of whites cultivating grand southern gardens. Read alongside representations of gardens by black and white southern women, "Old Subjects and New" represents one compulsory form of black and white togetherness in a new landscape order, a garden within a restored plantation. The photograph shows a brand of racial integration after Reconstruction, combining the black and white children through an allusion (to Stendhal's novel *Le Rouge et le Noir*) and the familial structure, for the children are "known to all the family" by the mock-pretentious literary nickname. The play on names is also racialized, however, as "Noir" names both Adeline's "dusky curls" and her social identity in the context of the southern garden. The picture's placement within the text suggests that it is

OLD SUBJECTS AND NEW IN AN OLD-TIME GARDEN
Note the pyramidal Boxwoods edging the entrance walk. In the foreground are Frances with her fiery tresses and Adeline of the dusky curls—known to all the family as "Rouge et Noir"

Photograph of H. C. Haynesworth's garden and home, Sumter, South Carolina. From Julia Lester Dillon, *The Blossom Circle of the Year in Southern Gardens* (1922), 42.

a model for the farms and country estates that Dillon hoped in her preface would rival those of any "section." That narrative of cultivation as regional progress has to be set alongside the story of black farmers in the South—whose numbers peaked in 1920—which has been described as "the story of justice pursued but never fully realized" (Green, Green, and Kleiner 60). Thus the picture is an ambiguous model for cultivation of southern gardens as a regional project of racial signification, landscape design, and even brand marketing. We are left to puzzle over whether the photograph is more hopeful or disturbing as a portent for the generation of black and white children who were of an age with its subjects.

A "Plain Dirt Gardener" and the Racial Imagination of Cultivation

In 1942, twenty years after the publication of *Blossom Circle of the Year*, a young professional landscape architect who wrote frequently for garden magazines, Elizabeth Lawrence, published *A Southern Garden*. Lawrence's manual was scaled for backyard gardens, rather than grand estates, across the South and lacked the ambitious reform agenda appended to the earlier book. Her friend Katharine White reviewed the 1967 reedition of the book in her *New Yorker* garden column: "*A Southern Garden* is far more than a regional book; it is civilized literature by a writer with a pure and lively style and a deep sense of beauty" (325). White singles out Lawrence's combination of precise horticultural knowledge with her clear prose and lack of sentimentality; such praise identifies by implication how *regionalist* was a pejorative term used to dismiss writers, ostensibly for sentimental place-attachment and limited perspective.

White's praise has also echoed in more recent assessments of Lawrence by historians and literary critics who have recovered, published, and edited her garden writing. The historian Vera Norwood and the literary critic Karen Cole, for instance, praised Lawrence's writing for its advocacy to preserve native plant species and its egalitarian correspondence with women across the South, including rural and urban women of varying classes and backgrounds. In her study of the role women have played in making U.S. environmental history, *Made from This Earth* (1993), Norwood in particular emphasizes how Lawrence, like Rachel Carson, Mary Austin, and other professional writers, scientists, and naturalists, shaped everyday attitudes through her writing and thereby contoured American landscapes. Cole argues that Lawrence "opened up the conception of the southern garden" (169) with her posthumous book *Gardening for Love* by attending to vernacular gardens. However, I find this argument more problematic when contextualizing the market bulletins that are the subject of that book.

Recuperating Lawrence's work in a national context, critics have neglected its complicated engagement in both the regional history of the South (including the imbrication of race and landscape) and the sometimes contradictory influences of literary and cultural "regionalism" as a position in debates over values or literary style versus "regionalism" as an approach to social scientific study. Southern regionalism in particular has a complex and, in the case of the prosegregationist Nashville Agrarians, disturbing history. Lawrence's more progressive views reflect the liberal approaches developed by the Chapel Hill social scientist Howard W. Odum, which she would have encountered in her studies in North Carolina. Lawrence's garden writing undoubtedly has influenced many gardeners across the South, particularly through strong club organizations that frequently invited her to lecture during her career (1930s–1970s) (Wilson 124). Lawrence's particular interest in the southern vernacular garden is better understood by connecting her work with more recent environmental and intellectual histories. Several strands of regionalism collide in her work: she advocated for native plants and heirloom cultivars that were disappearing from southern woods, represented gardening practices that crossed the color line, and emphasized farm women's cultivation through letter writing and plant exchange as essential to southern social history.

A Southern Garden is both a book of hours and a garden manual. As the latter, it deploys practical categories—botanical classification, a codified spectrum of colors—while relying on more implicit structures of nostalgia for what southern gardens used to be. In a section called "Daffodils in Old Gardens," Lawrence conjures an image drenched in place-attachment: "If you think back to quiet gardens in little towns passed by in modern times (you will have to go off the main highways to find them) you will remember the pale delicate pattern of the silver bells against dark cedar trees." Her reminiscence continues: "I have seen them in dooryards in Hillsboro, and in drifts under the oaks and along the box-bordered terraces of Coolemee plantation, and in gardens in the country" (30). The seemingly innocuous coincidence of design ("naturalized" white flowers on a dark background), nostalgic tone in relation to a plantation landscape, and later portraits of black gardeners from whom Lawrence bought plant material form a pattern of unconscious signification. This seeing and sorting in black and white sometimes would extend to the division of labor in garden making.

Lawrence was an educated professional writer, though she preferred a more modest self-presentation as a "plain dirt gardener" (*Gardens in Winter* 11). In 1932, she became the first woman to complete the landscape architecture course at what was then North Carolina State College in Raleigh. She began her career as a designer of local gardens and magazine writer before authoring *A Southern Garden, The Little Bulb, Gardens in Winter,* and *Lob's Wood,* as well as a regular column in the *Charlotte Observer* from 1959 until 1975. Lawrence corresponded with over two hundred gardeners across the country, and this work led to her final book project, *Gardening for Love: The Market Bulletins,* edited by the garden writer Allen Lacy and published posthumously in 1987.

A Southern Garden describes Lawrence's own garden in Raleigh (1930–40): it is organized by season and then by bloom sequence, categorized by family and genus. Like Dillon's *Blossom Circle of the Year,* Lawrence promises readers flowers adapted to the southern climate that will yield constant, coordinated bloom. Its geography of the South is fuzzy. The subtitle suggests that it is a "Manual for Gardeners in the Middle States," but the book incorporates letters from gardeners in Ohio, West Virginia, Pennsylvania, and as far afield as Scotland and England. Although she draws her botanical knowledge from a vast correspondence with contemporary gardeners and published authorities (her books offer a list of readings from Virgil to Gertrude Jekyll), Lawrence's scale of garden remains the modest backyard, measured in square feet rather than acres. The guide recommends color combinations for flowering borders and wild gardens that will work in a small space while suggesting the potential for enormous diversity of plants imported from similar climate regions that will thrive in southern gardens. The plants that she describes as growing in her garden are frequently memorials, signs of stories about the people who grew them, scions of friends' plants.[6]

A Southern Garden may be rooted in her Raleigh garden, but Lawrence weaves in stories excerpted from letters of friends, giving the names of gardeners and the cultivars they have grown with (or without) success. The book contains numerous indications of a secondary economy circulating plants and the names of gardeners as potential sources of the sought-after and rare. Outside the state botanical collection, the commercial catalog, and the garden center, Lawrence describes the circuit of informal exchanges familiar to passionate gardeners. Gardeners recur in these anecdotes, though they are identified not by name but by race and occupation; her named correspondents are racially unmarked. Black gardeners are described as "an old colored man" who sells door to door or the "colored woman" (43) who keeps a stall at the farmers market or the "laundress" (116) who is called one of "our colored friends" (57). Perhaps it is unsurprising the extent to which this secondary economy, which Lawrence might have been at pains to draw seamlessly into the seasonal blooming of *A Southern Garden,* emerges as moments of "local color," where color is also a racialized category evoking for her contemporary audience stereotypes and nostalgia for "old Southern gardens."

Narrating her own southern garden involves revealing traces of the racial division of labor that took part in its making. Read in the context of Jim Crow, when the "old colored man" brought redbuds "door to door" to provide foundation plantings for yards in her Raleigh neighborhood in the early 1940s, he may have been sent to the back door with trees to adorn front yards. To an extent, Lawrence's work as gardener and writer is captive to the function of her garden as a "model" and its circulation as a commodity. Her book was marketed with a photograph of the author posed within her Raleigh garden, embodying the iconography of southern white lady gardeners of leisure.

Lawrence actively performs the role of mistress elsewhere, as when she writes, "I begged from the laundress a start of the discarded *Tradescantia virginiana* (how often we go to our colored friends to replenish plants that they have kept and we have not), and planted it in a dark part of the copse" (57). Here, assuming a white audience ("we") also presupposes a degree of blindness to the economies regulating the circulation of plants and books. The historian Dianne Glave has explored the way women gardeners in the South worked within the constraints of a white-dominated economy, suggesting that something as everyday and easily overlooked as keeping older plant varieties, rather than replacing them with costly new ones, is coded by racial significance. Glave argues that "ethnicity was as important as gender in shaping the unique gardens of African Americans," in large measure because Euro-Americans have had more resources with which to purchase new plant material. These gardens "featured flowers, shrubs, trees, and plants that were purchased individually, accepted as gifts, or cultivated from cuttings," with "colorful motifs from gifts and cast-offs" (397). Gardens of relatively wealthier (on average) Euro-American neighbors might rely more on new materials and the latest fashions from gardening literature.

Glave's point is reinforced by an anecdote that Lawrence reports in *A Southern Garden*. The traces of racial division latent throughout the text come to the fore when she describes having purchased flowers from a woman at the market: "For years I struggled with hunnemannias [tulip poppies]. . . . Then one summer morning I came upon hunnemannias in a flower stall outside the market. They were fresh with dew, pale and delicately frilled, wreathed with fine green leaves, and altogether lovely. . . . The colored woman who kept the stall said she had grown them. I asked when the seeds were planted, and she said she couldn't remember rightly, but she thought it was at the end of March or the beginning of April. She had been cutting flowers for market since early June. 'They won't grow for me,' I said, 'I reckon colored people just grow flowers better than white people.' 'And I was saying to a lady yesterday,' she countered, 'that you all grow things better than we can . . . you have more to do with'" (116). Lawrence seems to have felt the justice of the poppy seller's complaint to the extent that she included the story as a fragment of dialogue. The positive stereotype, "colored people just grow flowers better than white people," offers insulting praise, which the market vendor counters with a cutting observation of racialized economic inequality. Reading *A Southern Garden* continues to be valuable because it exposes the dynamic by which the language of cultivation, the making of southern gardens, depended on work and knowledge within this asymmetrical racialized economy.

A Southern Garden further illustrates Toni Morrison's thesis in *Playing in the Dark:* that American literature has always had a strong Africanist presence, though one frequently suppressed, encoded, and stereotyped. The racial signification of Lawrence's book is in part a reflection of the historical context in which its author wrote and gardened: the segregated geography of Jim Crow–era North Carolina. The urban geography of Charlotte's neighborhoods was altered so profoundly in the postwar era that the urban historian Thomas W. Hanchett has described the process as the "sorting out" of the "New South City." The process of sorting out neighborhoods in Charlotte continued through the second half of the twentieth century, as indeed racial and economic segregation continue to reshape American cities.

The year she published *A Southern Garden* (1942), Lawrence and her mother moved to 348 Ridgwood Avenue in a new development on the edge of the Myers Park neighborhood, a white-collar suburb designed in the 1910s by the Progressive city planner John Nolen. Her garden is preserved today because the neighborhood was left intact, despite being the shortest path for a major street-widening project during the 1940s. The crosstown road, Independence Boulevard, was designed to curve around Myers Park, instead slicing straight through the African American neighborhood of Brooklyn in the southeast, erasing a good number of black-owned businesses and middle-class homes (Hanchett 262). Such a planning decision may explain why Charlotte, Lawrence's home city, was more segregated by 1970 than it had been in 1940, two years before she moved there from Raleigh.

The tense and divided urban landscape is a far cry from Dillon's holistic view of the South as a garden. Acknowledging this historical geography suggests that Lawrence's hopeful vision of a regional "Friendly Society" of letter-writing rural gardeners across the South, expressed in her final book, *Gardening for Love: The Market Bulletins,* is a limited utopia. *A Southern Garden* is symptomatic in its evocation of racialized divisions of labor in southern gardens, but the racialized imaginary and ideas of southern regionalism also situate what is resonant for theorizing cultivation and environmental justice in *Gardening for Love.* There, Lawrence explores the language and practices of gardeners, mostly women, who communicated through the market bulletins of the state agricultural agencies. *Gardening for Love* recuperates a hidden history of southern gardeners in the rural Deep South to preserve a "vernacular tradition" and to commemorate the major contribution of women to shaping American landscapes. Because that tradition crossed class and racial boundaries, Lawrence does not register the same worried awareness of the racialized division of labor as she did in *A Southern Garden.*

The epistolary form of Lawrence's garden writing reflects an earlier literary tradition of narrating complex events of domestic life in letters between intimate friends. The organization of her books resembles a latter-day eighteenth-century epistolary novel of gardeners' gossip. Yet there is another way in which this form's very roughness, with its fragments of letters embedded in paragraphs of description, literary allusions, and aesthetic advice gleaned from experience, suggests that Lawrence was not interested in using local geography to provide a narrative thread or organic whole. The transmission of letters in *Gardening for Love* also illustrates the variety of categories and ways of ordering garden knowledge. No single epistemology—historical, literary, or vernacular—could rein in the prolific potential of these southern gardens. Rather, taxonomies of bloom time and location (yard plants, box plants, window plants) proliferate alongside established ways of ordering plants. Even names of common species change across the region: the spiderwort (*Tradescantia virginiana*) Lawrence once had from her laundress she finds being traded in the *Mississippi Market*

Bulletin as "Texas bluebonnets" (71). Such heterogeneity in environmental knowledge across the region speaks to the area's botanical richness, climatic and soil variation, and historical ethnic diversity.

Yet Lawrence organizes this diverse knowledge in terms of the category "country people," echoing an early twentieth-century regionalist conceit that the "folk," in being closer to the land, preserved regional diversity long after urban modernization had homogenized and commodified a national culture. "Country people" became an evasive marker for Lawrence; she used it to refer to the "social history of the Deep South" recorded by the market bulletins. But in referring to country people, she also cites gardeners in rural areas outside the South as well as fictional gardeners, such as Sarah Orne Jewett's Mrs. Todd and her mother in Maine, with "her garden where the sea breezes were laden" (135). If early on her writing betrayed a racially charged nostalgia, her interest in southern gardens developed into a synthetic work of regional cultural study of the South's agricultural market bulletins.

Gardening for Love explores the lives and words of the "farm women" who became Lawrence's correspondents through the bulletins published by state agriculture departments throughout the twentieth century. As media of exchange, the bulletins were also agents of cultivation. In addition, Lawrence noted that they were "largely a southern institution" (26), one that resulted from regional economic development plans initiated by state governments along with other Progressive Era programs.[7] Lawrence was interested more in their literary qualities than in their context or institutional history, and her approach emphasizes the way the bulletins represented and bolstered solidarity among the "hard-working women" who "belong to that great fraternity, The Brothers of the Spade" (23). She signaled her interest in what the bulletins preserved: "times and customs that are long gone, or that are rapidly passing," the "poetic country names" for flowers and the "social history of the rural Deep South" that she suspected could be glimpsed in their pages (32). Lawrence praised the custom of exchanging news and expressions of care along with rare and disappearing plants: the phrase "gardening for love" names a moral vision of gardening as a social activity purified of selfish interests.

As a print medium of gardeners' conversations, agricultural bulletins were something short of a free exchange of plants for love. They could be as racially segregated as other modes of social interaction in the South and, indeed, were linked to institutions that historically excluded black farmers (Green, Green, and Kleiner 48). Lawrence overlooks the more distressing details in her rather nostalgic evocation of the personal ads; for example, offers for work and people seeking work frequently made whiteness a criterion of employment. So ordinary was this form of racial discrimination under Jim Crow that perhaps it seemed unremarkable to Lawrence when she began reading southern market bulletins in the 1940s, after her friend Eudora Welty introduced her to a sample from Mississippi.

Reading in the archive of such publications must remind us of how central race was to the social history they represent. For example, in the *Georgia Farmers Market Bulletin* of August 3, 1955, about half the "Farm Help Wanted" postings request white workers only; a few of the "Positions Wanted" postings advertise the aspiring worker's whiteness as a job qualification. Whiteness had a tangible currency, a calculable economic value in the exchanges taking place. No job seeker advertised being black. This competitive, racially stratified dimension disappeared from Lawrence's bucolic picture of the bulletins as a place where "farmers advertise their crops, their cattle, their horses and dogs, and their wives list the seeds and bulbs and plants that they sell for pin money." She described "whole families advertis[ing] for work on farms, and homeless people advertis[ing] for jobs with room and board" (35).

Lawrence read such advertisements for evidence of what she calls "the customs of the country people, their humor, and their way of speaking." She further framed her reading within the context of her admiration for Welty's fiction: "Like Eudora's novels, the market bulletins are a social history of the Deep South. Through them I know the farmers and their dogs, their horses and mules, and the pedigrees of their cattle" (35). Through this window into social history, Lawrence sought out recognizable types, imagining them as fictional characters and, moreover, coding even her social sympathies in literary terms: "Most of all I like to think about the hard-working farm women who are never too tired, when their farm work is done, to cultivate their flower gardens. They always find time to gather seeds, to dig and pack plants, and to send them off with friendly letters. To all parts of the country they send them off—yard plants, houseplants, and window plants. Reading the flower-lists is like reading poetry, for the flowers are called by their sweet country names, many of them belonging to Shakespeare and the Bible" (35–36).

Lawrence praised the market bulletins for creating a community of readers, classified them as poetry, and emphasized their power to disseminate gardens across the country. She imagined southern garden cultivation as multivalent: digging and gathering plants, raising cultivars, and elevating the cultural value of a community, in part by preserving links between vernacular culture (the "sweet country names") and high culture (Shakespeare and the Bible). Cultivation in and through the market bulletins became, in Lawrence's writing, an intensely regional work of preservation. It is unambiguously informed by literary versions of the South, in the way in which Lawrence's imaginative flight about the lives of the bulletin writers was clearly influenced by her own reading of Welty's fiction. Lawrence cites Welty's short story "The Wanderers," from the 1949 collection *The Golden Apples,* focusing on Miss Katie, "who advertised in the market bulletin" and sold from a roadside stand until "the wrong people took over the road, riding fast in their big trucks, hauling timber to mill" (36). The partial retelling emphasizes the importance that Lawrence attributed to the market bulletin as a synecdoche for an entire way of life that passed with the death of the fictional Katie Rainey.

In Welty's story, the tension between Miss Katie (Mrs. Fate) Rainey and her adult daughter Virgie is partly a generational economic shift: whereas the mother "used to set out yonder and sell muscadines" (525) and all manner of flowers in the market bulletins, her daughter "worked for the very people that were depleting the woods, Mr. Nesbitt's [logging] company" (516). Miss Katie's dying thoughts turn to the list of flowers, "her list" in the market bulletin, though she knows she is "about to forget the seasons, and the places things grew" (519). In Lawrence's recontextualization of this story, the market bulletin, its community of readers (country people), and the southern woods are all threatened with vanishing at the hands of those driving the logging trucks on the highway. "I think about the wrong people going by on the Mississippi roads and wonder how long the bulletin will continue to come," Lawrence writes (37). Bulletin readers, glimpsed through an elegiac short story, became her index for a regional social history that skirted the history of racial conflict in the Deep South but confronted the industrial creation of the New South.

In the final analysis, *Gardening for Love* claims not to be a complete social history, of course. Rather, it is presented as a deeply personal and creative investigation of attachment to people who represent both a region and a fictional lost past: "When I read the market bulletins I think about Mrs. Rainey and her flowers . . . when it arrives, twice a month, I turn the pages to see whether Mrs. Radau still has sunbonnet daisies and Voleene Martin still advertises the Texas bell vine. And then, when I find them all there . . . I feel comforted" (37). The peoples' names as much as the flowers' "old-timey" names are part of a moral economy that affirms order and continuity in the face of a rapidly changing society. The value that Lawrence finds in southern gardening, from *A Southern Garden* (1942) through *Gardening for Love* (1987), may be limited by its racial blind spots, but her complex pastoral critique of southern industrialization remains compelling.

The market bulletins are anachronisms that prove a rule of regional culture: even as it is marketed to outsiders, such culture has an aesthetic value (what Walter Benjamin called the "aura" of works of art) that perishes with its mass-production as commodity. The regional auras were particularly strong for individual plants kept in southern gardens that have since gone out of commercial circulation: "There are names not to be found in catalogs: the white coronation rose, the lady-of-the-lake, and the old fashion Betsy rose that blooms all summer" (Lawrence, *Gardening for Love* 38). Other species take on symbolic power, particularly as literary codes for race, by their very abundance and seeming autochthonous power, such as the magnolia and the chinaberry tree, which, although imported, seemed to spring up everywhere unplanted.

The generosity of Lawrence's correspondence with a society of gardeners across southern states, people willing to trade the plants, seeds, and letters that made possible such a remarkable book, holds out the promise of a different sort of regional environmental awareness. By her final book, Lawrence had turned toward a vision of vernacular gardens networked across regions. Given the dramatic social changes that were reshaping both rural southern areas and the geography of race relations after integration, it is perhaps unsurprising that Lawrence cast a nostalgic glance backward at what, for her, seemed to be a threatened vernacular tradition kept alive in the humble form of the agricultural bulletin.

Southern Garden Cultivation as Sacred Ecology in Alice Walker's Meridian

The writer Alice Walker has represented women's gardening in the South as a resistant aesthetic practice rooted in a desire for self-expression and beauty. Two of her earliest published texts pay close attention to valuing southern garden practices and reimagining their liberating potential: her first novel, *Meridian* (1972), and her often anthologized essay "In Search of Our Mothers' Gardens" (1974). Walker's vision of southern gardens appreciated their aesthetic quality and the sense of beauty embodied by rural African Americans and underpaid black sharecroppers working within an exploitative system. Like Lawrence's *Gardening for Love*, Walker's project was recuperative: she memorializes the unsung work of her mothers and grandmothers.

Despite sharing a general goal of recovering a hidden history of women's work, Walker and Lawrence diverge dramatically in the particularity of their chosen gardens. The South encapsulates multiple histories and several vernacular gardening traditions. From Lawrence's North Carolina we turn now to Walker's Georgia and Mississippi.

Reading the two authors side by side shows that seeds and plant material—the objects of gardeners' exchange—moved between black and white cultures late into the twentieth century, but carried with them different discursive meanings in these contexts. Walker writes of the gardens of black women's mothers in America, the places of beauty made by "all the young women—our mothers and grandmothers, *ourselves*—[that] have not perished in the wilderness" (235). By contrasting the space of gardens with "the wilderness," Walker refers to a central trope in American culture, familiar from a long line of scholarship from Roderick Nash forward. This mostly white, Euro-American genealogy tracks a progressive narrative from the "howling wilderness" of the Puritans to the "cathedral" wilds of John Muir, as Americans came to see wilderness as the country's most precious source of spiritual value. Yet in Walker's text, wilderness is symbolically linked with danger rather than sublime transcendence. This emphasis on the perils associated with wild landscapes in part reflects the experience of African Americans in the American South and elsewhere, where for many generations being caught far from one's community often meant the threat of harassment, enslavement, rape, or lynching. Ambivalence about wilderness has been expressed by many writers, beginning with what Michael Bennett describes as an "an anti-pastoral African American literary tradition" (195) inaugurated by Frederick Douglass, an idea elaborated eloquently by Evelyn White in her essay "Black Women and the Wilderness" (1995), in which White describes needing to work through feelings of intense terror, rather than blissful serenity, during her first visits to the deep woods of the Northwest.

Gardens have been a strong source of pride, identity, and meaningful work for women from the period of enslavement to the present. The historian Dianne Glave, in her 2010 study of African American environmental heritage, *Rooted in the Earth,* has emphasized not only the role of gardening as "preservation and conservation" for women in the rural South but also the rich variety of plantings and nonsymmetrical designs in their gardens (116). "By using yards, often in different ways than men," Glave argues, "women took possession of them. They manipulated and interpreted the spaces for sustenance, comfort, joy, and sometimes profit" (125). Glave based her research in part on recorded interviews with these gardeners, women similar to Alice Walker's mother, who gardened in Georgia during the 1930s. Walker's essay in 1974, published at the peak of popular wilderness environmentalism and protective legislation, anticipated the ways in which later writers have returned to everyday environments, urban spaces, and gardens specifically to explore ethical and political questions. Walker's essay in effect becomes a source text for emerging ideas of environmental justice.

Walker's search for mothers' gardens also articulates an ethic of writing—a set of beliefs about the obligations and responsibilities that writers, particularly black women writers, had to represent their mothers' stories. Walker found rich ethical value inherent in her mother's gardening—the way she grew flowers for beauty in defiance of her family's poverty. She created a caring environment and used all available resources to defy the crushing, oppressive economic conditions of the sharecropping system. Walker describes the flowers as adorning even the holes in their dilapidated cabin, an image that presents her mother's gardening as a defensive mechanism against a hostile socioeconomic environment.

Gardening, then, was not of ethical interest as a connection between the human and the nonhuman, but as an instance of spiritual work, a term Walker uses interchangeably with *artistic work*. The phrase has particular resonance for African American gardening traditions in the South, which are often linked to Christian and Central African spiritual practices of healing, protection, and community (Gundaker and McWillie 10). Black women's artistic work, Walker argues, necessarily defied unjust social relationships. The impulse to resist and challenge, rather than cave in silently, Walker calls the "secret" of her mother's garden, which is nothing less than a kind of psychic balm for what Wendell Berry has called the "hidden wound" of racism and environmental degradation that is particularly borne by southerners.

However, gardening for beauty and for the soul amid injustice was an ethical position that could be transposed into other aesthetic forms and activities. In this sense, Walker did not imagine gardening as significantly different from writing or quilting. When she describes an anonymous Alabama quilt maker using the "materials she could afford . . . the only medium her position in society allowed her to use," this work is set up in relation to her mother's efforts not only to encourage her children to read and write but also to make "ambitious gardens . . . with over fifty types of flowers" (241). The emphasis on flower gardening to the exclusion of vegetable gardening is also instructive. Walker places flower gardening in the column of art and soul; vegetable growing and canning presumably belongs in the column of necessity. This conceptual distinction might also reflect an abiding gendered and spatial division of labor for southern gardens, in which men more often managed row crops and vegetables for profit and women tended flower and kitchen gardens near the house.

A vegetable garden might feed Walker's family, but it did not offer significant material advantages or a way out of the exploitative economics of sharecropping. Seeing her mother enraptured by an activity not obligated by necessity, a compulsion sought and self-directed, was liberating. Subsistence gardening was still structured by the sharecropping system in southern Georgia in the 1930s and 1940s, so that whatever vegetable nourishment could be grown fed the workers' bodies only to the profit of the landowner. The land-rent agricultural system of sharecropping was based on a few landowners who extracted maximum profit by using tenants as a disposable means of production. Walker's vision of her mother's gardening militated against this systematic environmental injustice. Gardening elevated the gardener from the level of survival and necessity to a level of aesthetic choice. Walker says that in her garden, her mother became "involved in work her soul must have" (241), and this activity modeled strong self-possession for her daughter's generation.

Walker critiques socioenvironmental exploitation in the South throughout her protest novel *Meridian,* which is set in the waning years of the civil rights movement in Georgia and Mississippi. The radical cultivation represented in the novel goes deeper, imagining a residual cultural ecology for the region that resists both violence and complacency in the face of the status quo of white supremacy. The heroine, Meridian, is a student civil rights activist whose insights derive from contact with three enriching landscapes even as she internalizes the political conflict through periods of incapacitating and mysterious illness. Set sometime after the wave of assassinations and urban uprisings in the late 1960s, *Meridian* narrates the heroine's struggle to resist the violent legacies of institutionalized racism in the South while reclaiming images and stories rooted in southern landscapes as part of her full identity.

Walker weaves into the narrative three cultivated landscapes that have symbolic healing power for the narrator. The first is a mature magnolia, the Sojourner tree, on the campus of Meridian's school, Saxon College. The tree comes to stand for regenerative creative powers that precede the appropriation of the surrounding land for the university, whose name suggests it is a bastion of white cultural capital. Magnolias carry strong connotations in the American South; the Sojourner tree becomes a magnet for folk beliefs and other apocryphal tales. It is said to be a source of eloquence and musical genius for the students at Saxon, as well as an erotic rendezvous spot under its spreading branches. Walker roots the tree in the land's history of white brutality. Meridian and the other students repeat a legend that the tree's sacred powers come from its antebellum origins, when a plantation stood on the campus grounds. An enslaved girl, Louvinie, told her owner's children such terrifying stories that the man's son died of fright; as punishment, Louvinie's tongue was cut out at the root and buried on the spot, and from it grew the flowering magnolia. The Sojourner tree is thus presented as a parable not only of the psychic costs of white oppression on black memory but also of the legacies and displacement of racist violence. Walker describes Meridian as "holding on to" something that her friends, leftist radicals from New York, have let go to advocate violent opposition to white power (14). That something is her attachment to places like the Sojourner tree as well as to vernacular gardens and the rootedness and resilience they represent.

Early in the novel, Walker captures how this rootedness leads Meridian to view political violence with suspicion. Her fellow students riot on the Saxon College campus in response to yet another injustice, but "the only thing they managed to destroy was The Sojourner. Though Meridian begged them to dismantle the president's house instead, in a fury of confusion and frustration they worked all night, and chopped and sawed down, level to the ground, that mighty, ancient, sheltering music tree" (39). At a symbolic level, Walker resolves the tension between Meridian's commitment to political justice and her rootedness in the southern landscape by having the destroyed Sojourner send forth new growth by the novel's end.

Walker's Sojourner magnolia also represents a sacred tree, and indeed the other two landscapes that guide and heal Meridian are also linked to the sacred. The anthropologist Fikret Berkes has written of "sacred ecology," in which landscapes are tied to knowledge through long use. Berkes sees sacred ecology as typical of indigenous groups' knowledge of resources and relationships among organisms and natural forces, what has become a controversial discourse in anthropology and indigenous studies of traditional ecological knowledge. Walker clearly had an indigenous framework in mind, for the novel's epigraph from *Black Elk Speaks* prefigures the sacred tree that appears later in her text.[8] *Black Elk Speaks,* John G. Neihardt's 1932 poetic transcription of the Oglala Lakota chief's visions and life story, became a controversial "New Age" classic in 1960s counterculture. In the discourse of race, nation, and dead dreams in the post–civil rights era, perhaps Black Elk's vision crystallized for Walker the dangers of retrenchment and forgetting past struggles. The evocation of America's earlier history of racist political violence set against the culture-centering and sacred tree draws the reader's attention to two other hallowed sites in her novel: the great Serpent, a Native American mound on Meridian's father's farm, and the front-porch garden and "Kingdom" of Mrs. Mabel Turner.

The serpent effigy mound connects Meridian through deep time within the landscape. We learn how, as a child, she experienced a sensation of merging with the land while lying in a pit at the center of the coiled snake figure. The experience is made possible because her father, unlike his Euro-American neighbors, refuses to plow under the native sacred ground in the name of fencerow-to-fencerow profits. His empathy and conservation in turn resonate later in one of the book's turning points, when Meridian comes to identify with an elderly gardener, Mabel Turner. Meridian and her friend Lynne, a white woman from the Northeast, are registering voters in the deep country of Georgia and Mississippi. One of the elderly black women they approach is Mrs. Turner. The old woman's rural front-porch vegetable garden sets a scene that dramatizes interregional and interracial political tensions among the three women. As they approach, Mrs. Turner says she has heard about some "outside 'taters" in the neighborhood and then, becoming friendlier, accuses Meridian of "uprootin' my collards with yo' eyes." Meridian admits, "I sure was eyeball deep in her greens" (102). She invites the two hungry young women to dinner, and they accept, hoping to convince her to register to vote. Over an increasingly heated discussion between Lynne and their hostess, Meridian realizes that "Mrs. Turner was just well beyond the boundaries of politics" (103). This state of being is one that Meridian respects but Lynne disregards as the old woman's religious fatalism. Meridian, who has grown up on a rural farm in Georgia surrounded by pious Christians, realizes that living far from the institutions and conflicts of greater political powers, surrounded by her own garden, has insulated Mrs. Turner from formal politics.

Walker stages an argument between the white activist and Mrs. Turner until the old woman regrets having fed them her good butter beans and cornbread. What Lynne overlooks in her organizational, formal perspective on politics is the self-possession that Mrs. Turner has achieved, a union of soul food from her kitchen garden and her credo. Walker presents Mrs. Turner's garden as a kind of sacred ecology, a landscape of alternative relations that helped her escape some of the damages of the wider political struggle. Meridian's rejection of her friends' urban-focused aesthetic and political views results from attachment to southern regional ways, but the implication is that she is more aware of the connections of sacred ecology for her neighbors who have lived on and off the land for generations.

Meridian ends with the heroine's departure from both the civil rights movement and her friends, leaving "the sentence of bearing the conflict of her own soul . . . [to] be borne in terror by all the rest of them" (242). What does it mean to bear such a sentence? If my reading is correct, part of *Meridian*'s conflict is its divided attachment to southern gardens and landscapes. "Bearing this conflict" means dwelling with the question: How can southerners, white and black, care for landscapes that even in their most beautiful florescence bear the legacies of racial violence?

Gardening for Justice in the U.S. South as Global South

It is not enough to assume a gardener's view of the landscape without understanding the troubled past and present involved in making such gardens. The civil rights struggles at the center of Alice Walker's *Meridian* also fomented the community and land-based struggles of the environmental justice movement. That movement gained its momentum from communities of color in the American South who first resisted being targeted with waste incinerators and polluting industries in such places as Mossville, Louisiana, Sumter County in Alabama, Warren County in North Carolina, and the New South cities of Atlanta and Houston. The movement quickly grew and forged important interracial ties and alliances with religious groups through the 1980s and 1990s, its increased visibility aided by the scholarship of Robert Bullard, Barbara Allen, and many others since.

In the 2000s, the devastation of Hurricane Katrina, the Deepwater Horizon oil spill, and mountaintop-removal mining in the southern Appalachian Mountains reinforced a sense that southerners were bearing disproportionate environmental and social costs for economic benefits enjoyed elsewhere. This social geography of inequality has led critics to see the U.S. South as part of the Global South within the Global North, a vast region of resource extraction and human exploitation spanning the hemisphere.[9] While environmental justice in southern communities emerged first as an outcry against environmental racism, it has also become an organizing idea for political agitation in pursuit of better lives.

In the decades of accelerated global environmental change that followed the garden writing we have studied in New England and across the southern United States, questions of restoring diminished lands and healing brutalized peoples have only loomed larger. A long period of deindustrialization and growing social inequalities sometimes made public gardens flash points of resistance to what Rob Nixon has called the "slow violence" of extreme capitalist resource extraction meted out among the world's poor (32). Americans by the hundreds came to garden in the rubble of Detroit, a city wasted by financial speculation and loss of industry, and in the storm-razed acres of post-Katrina New Orleans. For the most part, these gardeners have been after something similar to the comfort, joy, and satisfaction that Mrs. Turner seemed to have found or that Lawrence imagined farm women across the South had cultivated in their gardens. Many have admittedly been lured and financed by corporate foundations to provide greenwashing for real estate development. But American gardens, and especially community gardens, also became symbolic and contested landscapes, suspect and utopian to their critics and sacred amenities to their advocates.

5 POSTINDUSTRIAL AMERICA AND THE RISE OF COMMUNITY GARDENS

The previous chapter explored how Alice Walker dramatized moments of intergenerational exchange within the garden settings of her novel *Meridian* and thereby raised questions about how we read landscapes that have violent legacies of racial injustice. In the late twentieth century, community gardens became ever more visible sites where this process of linking garden making and intergenerational landscape literacy could take place. The landscape architect and long-time community garden advocate Anne Whiston Spirn writes in *The Language of Landscape* (1998) that such gardens "create local heroes, leaders who become a source of advice and counsel for others with similar dreams. Adept at reading landscape, [these] community gardeners teach others such literacy" (212). Of course, the interdependence of community and garden means that cultivating such green spaces requires supporting the communities around them. As Laura Lawson observes in *City Bountiful* (2005): "Although garden programs seem to be perennial—appearing in times of crisis and disappearing in times of plenty—their constant reinvention begs the question of sustainability and the need for ongoing support. This support encompasses both the garden itself—the site and resources needed for plant growth—and the people who cultivate it" (xv). Yet when gardeners literate in the language of landscape and political power emerged in neighborhoods where their gardens stood on property valuable for speculative real estate developers, an explicit discourse of gardening and environmental justice developed.

In a satellite image of South Central Los Angeles from 2005, the blocks at East Forty-first and Alameda Streets stood out a deep green against the gray pattern of roads and roofs. For over a decade, that green rectangle of fourteen acres was the South Central Farm, a community garden cultivated by hundreds of Angelinos for food, community, and recreation. The farm also stood out against legacies of environmental racism that had made the area one of the most polluted in America. The image captures a point in time, a moment excised from the people and stories that sustained this particular place. By 2009, the parcel was listed by the city assessor's office as "vacant land," and a dun-brown, scraped field appeared in place of the lush gardens that once characterized the South Central Farm.

Comparing these before and after images underlines the fragility of community gardens; the stories behind such pictures help answer important questions concerning place making and the public function of garden writing: Why do some city neighborhoods nourish food gardens instead of urban parks? To the extent that stories about community gardens are "generic," what are their conventional episodes and endings? Finally, what powers, symbolic and material, do such stories have? Are community gardens in American cities really sponsors of environmental justice? Have they not also, as temporary welfare projects, sometimes served as a green front for urban real estate interests?

The South Central Farm seemed to many observers to be a parable of the oppositional meaning of community gardens in America writ large—those who saw it as a flashpoint of an emerging environmentalism of the poor, part of urban environmental movements reconnecting people and cities in late twentieth-century and early twenty-first-century America. Community gardens have been described as "patches of Eden" (Hynes), sites of community literacy (Spirn), "defiant gardens" (Helphand), and "insurgent public space" (Hou). In addition, social scientists, urban theorists, agronomists, and community activists have studied the community gardening movement as a source of community resilience and environmental justice, of capacity building and food security (Gottlieb and Fischer; Feenstra), offering increased social capital and economic benefits to surrounding property values.

The landscape architect Laura Dawson has to date written the most complete history of community gardens in America. In *City Bountiful*

(2005), Dawson traces the growth of the community garden movement to an era before the victory gardens of the First and Second World Wars and the vacant-lot programs during the interwar years. The City Beautiful movement of the 1890s encouraged school gardening schemes and relief gardens, which expanded through the influence of a new generation of urban planners during the Progressive Era. A few city gardens can claim to have community garden programs in operation for more than a century, such as those in Boston. Many cities began garden programs for unemployed, retired, or otherwise hungry and landless Americans in the 1890s, including in Philadelphia, New York, and Detroit, whose mayor, Hazen S. Pingree, turned vacant land into potato patches to feed unemployed workers. These gardens appeared in times of need and public demand, only to be built over in a pattern of bust-and-boom real estate development and urban economics over the course of the twentieth century. Dawson, with fellow architect and planner Jeffrey Hou, has also been at the forefront of translating research on the benefits of community gardening into long-term planning that includes such spaces within the urban fabric. Increasingly, their work can draw on the network of hundreds of gardens organized loosely through the American Community Garden Association, which has come to serve as a reservoir of data about the many benefits and planning alternatives offered by community-designed gardens.

Even more recently, filmmakers have examined the scope of community gardening as a movement of resistance to local political corruption and the monopolizing tendencies of neoliberalism. The producers Leila Conners and Mark MacInnis's 2011 documentary *Urban Roots* explores the way grassroots-led garden programs and more recent entrepreneurial urban farms have reshaped acres of Detroit in the wake of the city's long and then precipitous decline. The German filmmaker and landscape architect Ella von der Haide's four-part series *Eine andere Welt ist pflanzbar!* offers a global perspective of community gardening on four continents. Such movies reflect a plural, international land-based phenomenon that is too diverse to label as a single movement were it not for the increasing organization of thousands of gardeners worldwide into a global community.

Scott Hamilton Kennedy's documentary *The Garden* (2009) presented the South Central Farm as a parable of urban economic dispossession. *The Garden* adheres to patterns of environmental political documentary with a predictable cast of characters: charismatic *campesino* leaders, earnest civil rights lawyers, a self-righteous real estate developer, and even cameo appearances by sympathetic Hollywood actors and anarchist musicians. Despite its generic features, the film captures real cleavages within environmental justice activism. The site of the garden exists as a legacy of a successful struggle, championed by African American women at the head of Concerned Citizens of South Central Los Angeles, against a planned incinerator.[1] Starkly lit interviews set an office of Concerned Citizens present group leaders as morally suspect for compromising with city government and real estate interests. The film captures a political reality in which minority groups are played off one other and social movements seeking self-determination are tamed by political elites. But the narrative structure peddles a story of noble but ultimately defeated resistance to a stereotypical environmental villain: the property developer. The film thus functions as a "green screen" (Ingram 3) insofar as it obscures more complicated histories of conflict at the South Central Farm.

More ambivalent stories about the farm have dispersed across print ephemera, academic books and articles, and community websites, to which I will return at the end of the chapter. Although these community garden stories are fragmented and require work to reconstruct, a dominant narrative of urban development has since overwritten the site at Forty-first Street. That story—about the appropriate transformation of urban space—is the result of the interaction of government and property owners and relies on official discourse in politicians' public statements, municipal planning documents, city tax rolls, and a growing record of legal cases over contested garden sites.

In this chapter I situate and evaluate a diverse set of community garden stories within a cultural, political, and historical framework, analyzing narratives about the South Central Farm and New York City's community gardens while making excursions elsewhere, including to community gardens in Madison, Wisconsin. Collective narratives about community gardens are myriad and bounded in extrageneric ways by socioeconomic as well as biophysical environments. Through geographical, historical, and organizational diversity, it is possible to represent the plurality of community gardening in recent decades better than existing accounts, which have favored gardens' oppositional meanings and utopian promise while understating ambiguous cases or the appropriation of gardens for speculative urban real-estate development. The work of criticism cannot be ceded to satirists such as *The Onion*, which published an item in February 2008 that began: "Notorious for its abandoned buildings, industrial warehouses, and gray, dilapidated roads, Detroit's Warrendale neighborhood was miraculously revitalized this week by the installation of a single, three-by-four-foot plot of green space." Community gardening more often seems an earnest and earnestly needed enterprise and has tended to convert its theorists into advocates, perhaps for a number of reasons: the urgency of urban food access, unpredictable and rapid shifts in climate, and the growing economic inequalities in U.S. (and global) society.

Stories about community gardens contain multiple plot lines stemming from a variety of interests that cannot be subsumed easily by any singular drive, including a unitary public interest. As the interests of residents, gardeners, municipal governments, and private landowners interacted, at times in conflict and at times in coordination, two themes emerged in American community garden stories. First, land once more was the generative source of meaningful community, as seen in renewed civic discourse of homesteading, self-reliance, and national virtue, at times directly evoking the mass mobilization of victory gardens. And, second, land continued to signify the sacred boundaries of private property and a space of accumulation; but in a neoliberal political-economic context, urban real estate gathered around itself intensified symbolic powers.

By engaging these themes in critical ways, community garden stories have contributed to American environmental thought. As Anne Whiston Spirn and others since have argued, community gardens shaped and reflected community environmental literacy, a consciousness-raising about place and power that can occur at a collective level from neighborhoods to cities (212). In addition, these stories inspired debate over the status of food gardens on public property and then of the meaning of public space more generally in an era dominated by privatization and a growing gap between rich and poor. That debate at times spilled over into political action, policy, and legislation and led to broad claims to healthful food as a civil right, one critical to environmental and social justice in the contemporary world.

In these ways, stories about community gardens have offered a public deliberation of community and property in America. Such stories may not seem to fit neatly into the trajectory of twentieth-century garden writing mapped in earlier chapters, but they have reinvigorated discussion of the political and ethical character of gardening. One of the best contemporary writers on urban community gardening in America, Novella Carpenter, positioned her book *Farm City: The Education of an Urban Farmer* (2010) in a genealogy from the radical homesteading inspired by Scott and Helen Nearing in the 1960s through Michael Pollan's *Second Nature*—her book is one of the subjects of chapter 6, which looks at writing on gardens after 2001. Pollan also played a key role in fomenting this intergenerational exchange. He edited a Modern Library Gardening series featuring popular reprintings of classic British and American garden books and has since mentored newer writers, including Carpenter, in the school of journalism at the University of California–Berkeley.

It is widely understood that cultural narratives shape landscapes, but the reverse is also true. Dominant stories about democratic culture, land, and American gardening have indeed endowed community gardens with meaning, yet these gardens have likewise enriched the stories people could tell about themselves and the land. Urban community-garden stories continue a genre that the literary critic Michael Denning describes in *The Cultural Front* (1996) as the "ghetto pastoral," starting with proletarian city novels of the 1930s. These "tales of how *our* half lives" offered the "flip side of the slumming stories that ran from Stephen Crane to Dos Passos . . . not a proletarian sublime [as in Dos Passos] but a proletarian pastoral. These ghetto pastorals constituted a subaltern modernism and became the central literary form of the Popular Front" (231). New York's Lower East Side served as the urban landscape that charged the imagination of the proletarian writer Mike Gold at the century's beginning and garden resistance at its close. What Jane Jacobs called the legitimate desire of city dwellers to "dig in the ground" (144) motivated many citizens stuck in declining ghettos, where they nonetheless nurtured pastoral visions of better lives. These visions were often expressed and physically realized through gardens.

Gardens and garden-related organizations actively promote literacy of place analogously to other "sponsors of literacy," Deborah Brandt's term for "those agents who support or discourage literacy learning and development as ulterior motives in their own struggles for economic or political gain" (27). Community gardens are cultural constructions coextensive with other texts and practices of literacy, and in recent decades growing social-economic inequities have become involved in their making. This perspective on gardens is taken up by the poet, translator, and critic Jonathan Skinner in his discussion of ecopoetics, as well as by the French gardener, novelist, and philosopher Gilles Clément's idea of the "garden of resistance." In a 2011 essay, Skinner describes writers who "approach poetry as the extension, perhaps, of gardening by other means" (261) and ends with a question that reads like a political call to arms: "Who are the gardener-poets ready to join their tongues to these intermediate landscapes of the future?" (272). Some of these gardener-poets have already begun to create intermediate landscapes—between concrete and wild refuge—across American cities and claimed them for futures with greater social and environmental justice.

In the postwar period, economic policy and planning shaped cities and exurban spaces alike, sponsoring an uneven distribution of pollution and risk, which had a demonstrably racist dimension across regions and cities. The deindustrialized cities of America, from Detroit to the South Bronx, represented a vast swath of social wreckage that marginalized ethnic minorities and recent immigrant communities. It is no coincidence, then, that cities became seedbeds for community gardens and environmental justice activism after 1970.

Such activism addressed both negative (toxicity) and positive (amenity) dimensions of community functions and planning, but it was a product of two combined forces: the movement of environmentalist thinking sparked by Rachel Carson's *Silent Spring* in 1962 and the explorations of avant-garde artists in U.S. cities. The first community gardens began as public art happenings, first with San Francisco's People's Park in 1969 (Belasco 21) and then with a group of artists who in the early 1970s began cultivating gardens on abandoned, rubble-strewn, city-owned lots on New York's Lower East Side. As Liz Christy, a founding member of the Green Guerillas, describes in an interview, the group had grown fed up with the city's response to lots reverting to public ownership after being burned by owners for insurance money. The city had merely erected fences around the lots, which soon filled with rats and garbage. As early as 1972, residents broke through the fences, cleared the lots, and planted gardens. One of the first gardens, named for Christy, has survived and become a source of continuity and political meaning. The Green Guerillas, transformed into an official nonprofit organization, employ a discourse of environmental justice to advocate for gardens on behalf of marginalized communities. They in turn inspired recent groups worldwide to guerila garden by launching "seed bombs" into vacant lots, cutting fences, and planting trees and vegetables in neglected public spaces.

The number of community gardens nationwide grew slowly through the 1980s, concentrated in large cities with strong environmental movements, such as Portland, Seattle, and San Francisco, or large populations of immigrants from the rural South or overseas, including Hmong communities in the Midwest and African American and Puerto Rican neighborhoods in Philadelphia, New York, and Boston. Cities moved to fold squat gardens into public management, even when city parks departments were too strapped for funding to offer more than an official stamp of approval. As late as 1978, New York's official Green Thumb program, which required a symbolic one-dollar-a-year lease on existing guerilla gardens or risk a potential eviction notice, had no permanent staff. When the American Community Garden Association (ACGA) formed that same year, it represented a few dozen centers of garden organizing; two decades later, the group sponsored an annual national conference. By its thirty-fifth anniversary, the ACGA hosted a directory of thousands of gardens in North America on its website, supported research with Rutgers University on the multiple benefits of community gardens, and welcomed hundreds of members and international visitors to its annual meeting. Perhaps without intending it, community gardeners had become one of the largest environmental movements in the world.

In an essay reviewing the history of the environmental justice movement, the sociologist Robert Bullard tells a now familiar story of how the wilderness preservation focus of the environmentalist movement—dominated by the recreation interests of white, middle-class suburbanites—prompted minority activists to criticize the racist assumptions of mainstream organizations. This reading of environmental justice as an insurgency operating outside mainstream politics ought to be supplemented by recognizing the significant achievements made by coalitions that cross class, gender, and racial divisions, such as the community garden movement.

Urban parks have functioned ideologically, resolving social conflicts by formal means: they produced one illusion of nature as a well-tended garden while rendering gardeners invisible. In contrast, food gardens made by residents on dispersed vacant lots have brought people marginalized by urban planning to the frontlines of restrictive zoning and development plans. These community gardens struck directly at assumptions about urban parks. The latter work because they cannot be occupied permanently, and central authorities regulate their use, as witnessed in the wave of international demonstrations linked with the Occupy Wall Street movement after 2009. Like a demilitarized zone, large urban parks are calm as a result of arrested conflict; some have called Central Park the "eye of the storm" of the New York real estate market. The result has been conflict in some cities (New York, Los Angeles), accommodation and reform of park administration and planning to include food gardens in progressive cities (Seattle, Portland), and a revival of traditions of victory gardens elsewhere (Milwaukee, Boston).

Languages of Environmental Justice: Private Property and the Public Interest in New York's Community Gardens

In cities dominated by neoliberal reforms (defunding of public services, privatization of tax revenues as business incentives) through the 1990s, there was a growing sense that food gardens might be more than a weird agrarian anachronism or "anachorisms," elements out of place (Cresswell 166). Despite, or even because of, the force of the dominant story about land as private property, community gardens have proliferated in cities. This fact led the landscape architect Kenneth Helphand to designate them as a recent form of "defiant gardens." In his 2006 book of the same name, Helphand provocatively links urban gardens to other gardens made in extreme conditions: in the trenches of the First World War and in Polish ghettos, internment camps, and POW camps during the Second World War. The term defiance may be an allusion to the essay "On the Defiance of Gardeners" by Henry Mitchell, whose Earthman persona we encountered in chapter 3 in connection with the garden-centered ethics of Michael Pollan. Mitchell attributed defiance to gardeners who face annual droughts, freezes, and pests yet return each year to plant. Helphand's gardeners faced the most adverse conditions imaginable of modern war and dispossession, and he takes them to exemplify a universal impulse, an urge to cultivate for food and beauty. Whereas Mitchell's "Defiance of Gardeners" considers nature the gardener's sole adversary, Helphand's gardeners also defied the social injustices that shaped their impoverished condition.

Evicting gardeners from public land asserted the dominant narrative about American cities as spaces of capital accumulation, which is the story most often reinforced by government authority. This view projects an imagined future in which a developer, abetted by generous bankers, planners, and local government, eventually reverses the effects of economic downturn in the last three decades and creates jobs for unemployed residents, perhaps private schools for every child, and swaths of lawn and recreational wilderness. Such cities of the future imagined by neoliberal policies may be improbable, yet in the words of the former New York City mayor Rudy Giuliani, a powerful booster of this vision, it is community gardeners who "live in an unrealistic world," not speculative developers (Chivers). Greenspace is not exiled from the technocratic developer's vision. In fact, the utopia variously referred to as the "postindustrial," "postmodern," or "global city," or by its critics as the "neoliberal city," represents a kind of urban pastoral, distinct from the bottom-up view of ghetto pastoral narratives. From Central Park to the Meadowlands, compelling evidence shows that urban parks can and do function as a green front for real estate development interests. "The park," writes Robert Fitch in his explosive history of New York City, "is thus the modern planning equivalent of the medieval moat, protecting the barons' castles" (71). Moreover, Fitch argues that several of the larger public parks in Manhattan not only prevented working-class neighborhoods and industry from expanding but also bolstered discriminatory residential values.

In New York, a group that became the Environmental Justice Alliance (EJA) played a pivotal role, articulating theoretical claims about environmental justice in relation to community gardens and contesting public and green spaces. The EJA sought to defend community gardens across the city by bringing suit against the mayor when the city sold garden lots to developers during the rising real estate market of the 1990s. Their high-profile struggle represents an environmental justice praxis in which urban gardeners across the country have claimed community gardens as instruments and representational spaces of justice. Nationwide, gardeners dramatized a connection between environmental justice and food gardens through legal contests, writing, public performance, guerilla theater, installation art, and civil disobedience.

Community gardens as environmental justice praxis supplement in imaginative ways the shortcomings of procedural justice. The critic Wai-Chee Dimock has called the "residues of justice" the claims that cannot be resolved neatly by a legal framework that instead inhere in writing that engages injustice thoughtfully within its historical context. Dimock contrasts "two primary languages of justice, law and philosophy" with "an alternative language, the language of literature," claiming that imaginative writing records the incommensurable claims to justice that are only ever partially adjudicated by legal decisions and philosophical concepts of justice (8). Dimock calls this alternative language "literature" but also considers it a cognitive mode, asserting that language in this mode represents justice in ways that are more hopeful and less overbearing than the "dream of objective adequation which justice is." She asserts that "literature" as a "semantics of justice" is thus "the very domain of the incommensurate . . . of the nonintegral . . . it denies us the promise extended by law and philosophy. . . . For that very reason it is a testing ground no jurist or philosopher can afford to ignore" (10).

Following Dimock's suggestion that literature is an alternative language as well as a cognitive mode, it makes sense that stories about community gardens also preserve "residues of justice," that is, remnants of two prevalent beliefs about the commensurability of legal justice. First, that legal language can adequately balance protection of public land for public use in a culture dominated by the public influence of private property (in the form of real estate interests). Second, that existing legal institutions can identify and arbitrate a single public interest amid a multicultural community.

The incommensurability of these goals is evident in prominent legal rulings on community gardens, which have been found to be negligible uses of public land if local government seeks to convert such sites into private property (as occurred in New York City when the mayor sold garden lots to private housing developers). Outside the legal sphere, nonprofit institutions, scholars, and activists supported community gardens by claiming that they redress institutional racism and the socioeconomic disempowerment of American ghettos. Activists and gardeners writing first-person accounts of lost gardens have not claimed that specific community gardens fully "balanced" or "compensated" for past injustices. Rather, they see gardens as long-term projects working toward justice. This alternative language of community gardens was anticipated by earlier garden writing and other versions of the ghetto pastoral. It existed, too, in the themes of land-based struggle found in Native American literature, which the literary critic Joni Adamson has shown offer many of the clearest and earliest framings of environmental justice in North America.

The most significant finding of legal proceedings is that, as public property, community gardens do not provide a legal basis (i.e., standing) for disputing their loss as a violation of the civil rights of individual gardeners. For community gardens on city land, procedural justice is the language that has power to articulate the boundaries of public and private in America. It is the language backed by state force, empowered to transform public spaces in cities. Community gardens on public land enter this language by virtue of their symbolic power, by appealing to cherished ideas in U.S. culture about agrarian democracy and self-reliance. Still, the assumption is that food gardens ought to be on privately owned land; even political radicals like the Nearings (in their homestead) maintained the dominant view of edible gardens as inherently private. Public parks in America have functioned as an expansion of the private space of a particular class. One reason for the unstable legal status of community gardens is that they defy standard categorization of public and private land. In American culture we experience a bifurcated discourse of gardens—on the one hand, a legal and philosophical language of public parks, green space, and wilderness; on the other, a tradition of garden writing that has cleaved to private and domestic spaces, as examined in previous chapters. The keyword in foregrounding this relation of people and public space is *public interest*, the crux of urban planning.

When public land is at stake, decisions turn on the interpretation of the public interest by residents, landowners, and courts. A persuasive—yet hardly monolithic—meaning of the *public interest* must be established to change the legal status of public land and the permissible range of its uses. A legal construction of the public interest does not police all the ways that public land is used. A single city block witnesses a stunning plurality of interests, including those of people marginalized by urban development schemes. Yet city dwellers who appropriated empty public land are out of place in a legal setting without a notion of usufruct or tradition of commons. The cultural phenomenon and legal institutions reflect two distinct classes of interest and cognitive modes, one in which a justice system defends individual property owners and defines food production in market terms, segregating it from a "cognizable" public interest, and another that seeks environmental justice by recognizing links between the social history and biophysical cycles of communities of people and plants.

In the case of community gardens in New York, stakeholders have bargained within existing property regimes to continue to plant each season. This has meant accepting temporary leases on public land through the city's Green Thumb program and, more recently, finding a compromise between private and public ownership. In the latter arrangement, several private nonprofits (notably the Trust for Public Land) have purchased gardens on city land that was scheduled for housing development, securing the gardeners' tenure. Community gardens have endured as a political compromise between public and private ownership, that is, as exemplary spaces and cautionary tales about the historical and political limits to realizing environmental justice in U.S. cities.

From this perspective, community gardens are notable for how they mediate the contradictions between private property regimes that shape urban economic geography in relatively rigid, uneven ways and a civic ideology of cities as places of democratic exchange, cultural freedom, and economic mobility. One of the most significant initiatives that makes visible the democratic, symbolic power of community gardens is the New York City–wide Community Garden Mapping Project. The Council on the Environment of New York City, a private organization, began the project as part of its Open Space Greening Program (OASIS). Web-based GIS technology allows visitors to produce many-layered maps of city greenspace: each represents how the city "sees" a lot (in tax and zoning terms) as well as what is actually there (as a brief history of its use). Community gardens appear as clusters of light-green dots in areas with relatively fewer public parks, represented by darker-green areas. The project relies on local gardeners and groups to update information from the ground up. OASIS offers eloquent collective images of community gardens that establish equal footing between different types of public parks and represents a direct mediation of public interests and private property, insofar as the maps offer a visual counterpoint to Mayor Giuliani's decision in 1998 to move Green Thumb gardens from the parks department to the city housing agency.

The city, as custodian of over ten thousand vacant public lots, approached the existence of community gardens with a range of attitudes, from open hostility toward gardeners who defied Green Thumb recognition to tacit acceptance and even celebration of particular gardens. Significantly, for years the cultivated lots remained "vacant" on paper even when Green Thumb leases had been granted to establish gardens in them. They were zeros in the tax revenue column of the city budget. In April 1998, the mayor's office shuffled many officially empty lots from the Department of Parks and Recreation to the Housing Preservation and Development (HPD) agency. The bureaucratic shuffle redefined the gardens' meaning, from blooming community green space to blank potential real estate in a rising market. Over one hundred such gardened lots went up for auction. By 2000, ninety-one had been destroyed, including thirty-two in Harlem and the Lower East Side alone (Von Hassell 162). Many gardeners responded to bulldozers with direct action and political pressure; legal claims relied on articulating a discernible public interest within the justice system. The resulting legal record again represents a horizon or limit to interpretation of community gardens as public spaces.

The legal battle between the Giuliani administration and the state's attorney general, Eliot Spitzer, epitomizes the justice system's role in according legitimacy to certain representations of gardens. The representations of community gardens in legal discourse sorted a complicated cultural narrative into, on the one hand, a series of legally cognizable interests defensible with existing statutes and constitutional law and, on the other, incoherent claims and inadmissible evidence. The city's decision to sell garden lots to housing developers led to a court battle between Mayor Giuliani and a coalition of garden activists represented by the New York City Coalition for the Preservation of Gardens (NYCCPG), which in turn became the New York City Environmental Justice Alliance (NYCEJA). The judge in the initial case in 1997 asserted that the city represented the only "legally cognizant" public interest in question, given the status of the lots as public land leased to private individuals (NYCCPG v. Giuliani). The city claimed to represent the public interest by promising to build affordable housing; the NYCEJA subsequently claimed that destroying existing gardens on public land violated minority gardeners' Fourteenth Amendment rights (NYCEJA v. Giuliani). The state's court justices repeatedly rejected the NYCEJA suit. Gardeners renting public land did not have legal standing to complain that selling the land violated their rights—as nonowners, they had no particular rights to use the land. This opinion was upheld on appeal, affirming the denial of community gardeners' standing. Sale of lots and bulldozing continued until 2000, when Attorney General Spitzer succeeded in winning an injunction based on an argument that losing the gardens would cause residents "irreparable harm" (State v. City of New York). The injunction forced a settlement in 2002, when the attorney general bargained with a new city administration under Mayor Bloomberg to create a procedure for garden review and a process for legalizing community gardens that have obtained Green Thumb status. The government as owner of public property remains sole legal representative of the public interest.

In practical terms, the power of local government to determine the use of public property means that officeholders who advocate privatization are empowered to represent this agenda as the public interest. In the case of Mayor Giuliani's administration, the city clearly advocated privatization, reflecting a belief that speculative building and gentrification were better indices of neighborhood development than were community gardens. Privatizing community gardens once on city land proved a successful strategy to preserve them in the context of a neoliberal regime. As a representative of the Bronx United Gardeners has noted, "privatization [became] an effective way to make gardens permanent" (Woefle-Erskine 54). The legal rulings over New York City's gardens in the 1990s, which repeatedly denied that food gardening was a "legally cognizable" use of public property, effectively supported privatization. Yet the mechanism of privatization, and perhaps its very meaning, is challenged by the nature of community gardens in American cities: civil society, in the form of groups like the Trust for Public Land, widened the meaning of *public property* by purchasing public land and keeping it available for public use in modified forms, including for food gardens. Where legal language forcibly equating public property and public interest has rendered community gardens and gardeners mute and invisible, other forms of writing have explored the meaningful dimensions of community gardens.

Bourgeois Fantasy and Ghetto Pastoral

Experiences of dislocation and dispossession characteristic of late twentieth-century urban communities can be difficult to narrate, which perhaps explains the relative absence of community gardens in fictional writings. The cultural historian Michael Denning studied the resistance of working people's experience to narrative and coined the term ghetto pastoral to describe a form of proletarian writing for "figuring an imaginative space, a place of poverty and lack" that emerged in the 1930s (244). Throughout this chapter and the next, I will use the category of ghetto pastoral to theorize connections between environmental justice, bourgeois fantasy and gentrification, and real-world urban gardens.

The ghetto pastoral and its "dialectic of degradation and elevation, the grotesque and the simple" (251), has been infused not only into the wider culture, as Denning argues, but more specifically into the fragmented structure of postmodern novels. In his prize-winning 2001 novel *The Corrections*, Jonathan Franzen stages an imaginary community garden at the center of the book, in a section titled "The Generator." Franzen's depiction of a so-called Garden Project in Philadelphia unearths the ideological as well as the critical work of middle-class fantasies about community gardens as rectifying the unfairness of America's declining minority neighborhoods. Christian Long has noted Franzen's interest with the geography of suburban–urban relations, presented in scenes of commuting in the fictional midwestern city St. Jude. Read against real maps of St. Louis, which St. Jude most closely resembles, Long finds that the book offers a cognitive mapping of white middle-class psychology (210). I argue that the fictional community garden also functions as a fantasy setting for the affluent white characters, who initiate and manage the project on behalf of African American residents and garden workers. Although not the first parody of a community garden—the novelist David Lodge satirized the Berkeley People's Park, created in 1969, in his 1974 novel *Changing Places*—Franzen's is more sustained and systematic. He presents the "Garden Project" potentially as simultaneously a utopian space, a revival of ghetto pastoral, and a project of white gentrification and bourgeois fantasy.

Throughout "The Generator," Franzen develops several critical dimensions of American ghetto pastoral, thematizing the relation of urban community gardens, class, race, sexuality, and narrative form. If we judge by Franzen's own statements, *The Corrections* represents a conscious attempt to continue a tradition of novels as a form of social critique. In a widely read *Harper's* essay of 1997 ("Perchance to Dream," reissued in 2002 as "Why Bother?"), Franzen advocated a brand of social realism informed by postmodern style that resists empty irony. *The Corrections* seems to deliver on that promise, offering a sustained critique of postwar American capitalism, from industrial consolidation to speculation, peaking with the late 1990s boom and bust. Franzen deftly orchestrates the crisis of American political economy as it works through the psyches of the Lamberts, mapping financial swings onto the psychological disintegration and transformation of Alfred and his wife, Enid.

The story, set in a Philadelphia ghetto, follows Denise, the youngest of the three adult Lambert children who rises to national acclaim as the culinary genius behind the Generator, a chic restaurant built within a former coal-powered electric generator. It is Denise's affair with Robin Passafaro, who directs the Garden Project, that provides the novel's galvanic force. Robin is married to a dot-com entrepreneur named Brian Callahan, who funds the Generator to put his hometown "on the map of cool," as a fictional restaurant reviewer says. Plagued by guilt over her husband's sudden wealth, Robin throws her energy into cultivating the community garden on an empty lot in a gentrifying neighborhood a mile from their street.

The Garden Project is at once a hapless gesture toward redistributive justice, a way for Robin to assuage liberal guilt at buying up space from marginalized black Philadelphians, and an erotic Eden in which Denise and Robin realize their desires. That the garden is transplanted into a community becomes clear as soon as Robin introduces Denise to the project, which occurs in the scene when they first become friends. Robin gives Denise a tour of the recently cleared field, admits that the project is "selfish" and stems from a childhood dream to "have a big garden," which she calls "her enchanted kingdom" (403). Robin staffs her garden with African American teenagers from the neighborhood. Franzen thereby raises the notion that such charity gardens are in fact a form of neocolonialism: Robin uses her husband's dot-com wealth to hire cheap seasonal workers and realize her private garden fantasies, maintaining a black underclass in an otherwise economically marginal ghetto. Brian's first purchase is a large house on Panama Street, another detail that marks the territory as a microcosm of American imperial projects. For all her sensitivity and family roots in leftist politics, the fictional Robin's enthusiasm for the Garden Project resembles racial paternalism. Franzen's ambitions (and ambivalence) regarding the politics of the novel, which exploded in a public falling-out and subsequent apology to Oprah Winfrey after she chose *The Corrections* for her book club, make his representation of a community garden project as bourgeois fantasy all the more provocative.

The history of Philadelphia's community gardens is more complicated than white paternal fantasy and comes closer to a revival of the insurgent force of ghetto pastoral, that vision of a more just and transformed America born in the imagination of 1930s proletarian writers. African American community leaders organized many of Philadelphia's community gardens. Several leaders drew on extensive horticultural knowledge from their childhood in the rural South, effectively transporting residual ideas of gardening and community when they migrated to northern cities such as New York and Philadelphia (Hynes 11). Others were genuine interracial collaborations that crossed lines of class and education, such as the West Philadelphia Landscape Project. Fiction need not reflect history, of course, yet the comparison raises questions about the susceptibility of depictions of community gardens to what Leo Marx called a naive or "simple pastoral." Such pastoral art is innocent of historical awareness of landscape transformations and real environmental and social damage; aesthetically, it is free of irony.

Franzen's Garden Project certainly reveals an irony in Robin's attitudes and behaviors—a situational irony shared by affluent newcomers who advocated for community gardens in "their" new neighborhoods while participating in reshaping the economics of the place, driving gentrification that priced out earlier residents. The Garden Project also marks Robin's practical attempt at environmental justice via redistributing the use and profit of urban land to black residents. To neighborhood high schoolers she offers free horticultural instruction as well as a share of vegetable sales in exchange for a modest amount of physical work: clearing weeds, digging beds, planting seeds. Nor does she draw a division of labor between herself and the other gardeners. Denise first spies Robin as a figure in the distance, clearing stones long after her young helpers have left.

"The Generator" also narrates Denise's developing sense of her sexuality—as a lover of women who remains deeply invested in heteronormative shaming of lesbians—in ways that recognize a tension in asserting an identity without being narrowly defined by it. The community garden also becomes the place where Denise and Robin seem best able to act out their mutual desire. Denise Lambert enters the Callahan-Passafaro household at the destabilized moment when Robin has just begun the garden, which becomes their erotic rendezvous. Here Franzen trades on the liminal quality of city community gardens. His characters are revived in a memorable tryst that leads the reader from an allegorical urban landscape through a series of increasingly corporeal images to a climactic sequence of protests:

> She drove through Crack Haven and down Junk Row and past Blunt Alley to the Garden Project, where Robin had a blanket. Most of the garden was mulched and limed and planted now. Tomatoes had grown up inside bald tires outfitted with cylinders of gutter screen. And the searchlights and wing lights of landing jets, and the smog-stunted constellations, and the radium glow from the watch glass of Veterans Stadium, and the heat lightning over Tinicum, and the moon to which filthy Camden had given hepatitis as it rose, all these compromised urban lights were reflected in the skins of adolescent eggplants, young peppers and cukes and sweet corn, pubescent cantaloupes. Denise, naked in the middle of the city, rolled off the blanket into night-cool dirt, a sandy loam, freshly turned. She rested a cheek in it, pushed her Robiny fingers down into it.
>
> "God, stop, stop," Robin squeaked, "that's our new lettuce." (418)

Following this climax in a private corner of the garden, Franzen unknots the lovers by having Robin cut off the affair, which prompts a vengeful Denise to seduce Robin's husband. The affair, like the strangely anthropomorphized plot of "pubescent cantaloupes" and "adolescent eggplants" where the lovers bed down, fades from *The Corrections* without a glimpse of a harvest. The ironic depiction of Robin's Garden Project as a bourgeois fantasy, able to make temporary room for sexual adventures, undercuts the novel's otherwise conservative thematic interest in marriage, family, property, and identity, explored in the longer sections that follow Denise's brothers Chip and Gary. The symbolic debunking of novelistic narrative in "The Generator" incorporates community gardens as a trope but suggests that this fictional garden project is no more edifying than the broader fictions of gentrification: that somehow ailing cities can be renewed by replacing the people in a community with different people, that the fictional classlessness of suburbs can remigrate into gentrifying "mixed neighborhoods" and become havens of greater diversity than the normative suburbs.

Ghetto pastoral thus describes the literary form and political meaning of the Garden Project in *The Corrections*. Yet, taking a cue from Franzen's well-known statements about the crisis of contemporary American fiction, I also want to identify a limit on the political efficacy of such work. In Franzen's "Why Bother?" essay, he describes an anxiety that fiction with a grand social agenda is unlikely to find an audience in contemporary America. He accounts for his personal resolution to write fiction rather than political criticism in part as acceptance of a division of labor: according to Franzen, nonfiction simply reports social problems more effectively (66). The critic Robert McClaughlin is mostly right in grouping Franzen's work with what he calls "post-postmodernism"; insofar as *The Corrections* offers social-historical context as counterpoint to each of the protagonists' lives, Franzen's "response is . . . essentially conservative" (61). I would argue, however, that Franzen's critical statements do not track with his fictional creations. "The Generator" section of the novel flirts with engaging specific, historical problems of environmental justice in contemporary American cities. At some level this engagement is diminished by the episodic narration of the Garden Project: its energies are exhausted along with Robin's guilt and Denise's thwarted desire, and the romantic plot reduces the community garden to a backdrop. As a novel, *The Corrections* is fleetingly interested in the subjectivity of Robin and Denise, with their projects, the Generator and the garden, illuminating (like another of his "compromised urban lights") a falling trajectory for their erotic relationship.

Adopting the ghetto pastoral as a cultural narrative is problematic because its utopian hopes for neighborhood redevelopment seem to come with the downsides of gentrification. On the ground, gentrification often renews ailing neighborhoods by replacing poorer residents with wealthier newcomers. Community gardens run the risk of ratifying gentrification of working-class and ethnic communities by green-washing white guilt and real estate speculation while transforming ghetto rubble into private kingdoms. From this perspective, Robin's fictional Garden Project resembles promotional advertisements for chic upscale urban communities.

Nonfictional stories about New York City's community gardens in the 1990s at times functioned as an ideology of gentrification. For example, Robert Fitch's unveiling of the role of real-estate-funded charitable foundations in forming urban gardens in the South Bronx and Brooklyn indicates that ghetto pastoral tropes were sometimes coopted by privatization development agendas. The Lower East Side community garden skirmishes of the 1990s followed an earlier city campaign to clear homeless squatters from public parks. Everyday problems and networks of jobs, relationships, energy, and resources can also disappear behind the pastoral visions of community gardening. By overstating the symbolic power of community gardens, we further run the risk of exaggerating their political efficacy to foster economic and social justice.

The Symbolic Power of Marginal Gardens for Environmental Justice

The work of cultivation in urban community gardens pushes gardeners and observers to the edges of such literary forms as ethnography and participatory journalism. To dig deeper into the symbolic power of urban community gardens in cultural narratives, let's consider *Transitory Gardens, Uprooted Lives* (1993), a remarkable photo essay by the architect–photographer team Diana Balmori and Margaret Morton documenting the most marginal of New York City gardens. The visual dimension of Balmori and Morton's work lends itself to exposing the dynamic of visibility and power in stories about community gardens, which helps us imagine a just environment not only because of the unique situation of the gardens, but also because of the problem of seeing and representing them. Where one person sees a garden, someone else may see a symptom of blight or merely a vacant lot. Balmori and Morton present multiple visual and narrative angles on gardens that are hard to see, what they call "transitory gardens" that were made by New York's homeless in the 1980s and 1990s. In Balmori's text and Morton's photographs, later community gardens appear as more permanent and stable versions within a typology of marginalized green spaces that the authors developed in *Transitory Gardens*.

Reading Balmori's interviews with transient gardeners can help us better understand the way property and liberal individualism often have underwritten the work of gardening—even when community gardens have made compelling claims to represent more just social environments. The lavishly illustrated, high-production book points us in the direction of this problem of seeing marginal gardens in the city. By chronicling gardens made by squatters and artist-activists as well as more established community examples, *Transitory Gardens* provides a necessary counterpoint: these have been spaces of conflict as well as community, reappropriation, and forgetting.

Balmori and Morton begin with a revealing categorization of urban gardens and then describe how the Green Thumb program coopted lots in the 1990s that had been recently cleared of homeless residents. Such community gardens were political creations, the result of police force. The Eighth Street garden is a revealing example: the Green Thumb program encouraged a community garden on a plot cleared of squatters months before. Balmori and Morton recorded the displaced residents' stories, including visual and textual accounts of their gardens.

One black-and-white photograph features Pixie, a homeless woman who lived on the Eighth Street lot. It shows her garden, made in the absence of access to soil and water, shortly before police drove her away. A potted flower, a green carpet, chairs, and an umbrella compose her self-described sitting garden. In an image on the next page, we look back through the fence into approximately the same place occupied by Pixie's garden several months earlier, now a community garden. The juxtaposition reveals how a community garden was built from the litter of dispossession; Balmori draws our attention to where one gardener used the police line barricade as part of a raised bed. Such images expose how the stories of gardeners, particularly those created by the homeless, enter into published discussions only through aesthetic translation. Clever and desperate garden-making is translated into elegy by sequencing these before-and-after images.

The elegiac quality of many of the photographs complicates the testimony that Balmori claims the gardens make on behalf of human needs: "The gardens illustrated here, pared of the superfluous, made with true economy of means by persons who are deprived of the most basic necessities, seem to point to the power of the garden laid bare. To us they seem a testimony to the essential need for a garden: perhaps through them, we can learn again what a garden is" (7). The authors grant an audience and a hearing to a testimony of homeless and squatter gardeners. These gardeners had neither legal standing nor political allies, and so they were evicted and otherwise rendered invisible in the slum clearance of the 1990s. Ironically, their inability to prevent police from razing their gardens is presented to the reader as the symbolic power of gardens: "the power of the garden laid bare."

This power is also the power of looking that we exercise collectively when figuratively uncovering the sense and meaning of such gardens. We might charge Morton and Balmori with aestheticizing victims, making an artful exposé of New York's street life in the tradition of Jacob Riis's *How the Other Half Lives*. At times Balmori's text invests too much in aesthetic, symbolic power, as when it reports the destruction of another garden featured in Morton's pictures: "Jimmy's garden was bulldozed about eight days after this photo was taken, and he has disappeared from the area. But the moment in the garden that this picture captured remains—in this book and in him" (62). Is it valid to identify that "moment in the garden" as one shared by Jimmy and the viewer, across all other partings of difference? Just what *does* the image capture? That moment is not exactly shared, though juxtaposing the portrait of Jimmy against his razed garden invites readers to empathize with his loss and, by viewing the contrasting scene of a clutter of broken concrete, to appreciate better the intentional design of his garden. That garden included a pond lined with waste plastic that incorporated used tires and other refuse, producing a visual effect reminiscent of both vernacular art and rural gardens, where bottle trees, rusted machine parts, and tires spray-painted white or metallic gold are not uncommon.

The sequence reveals a history of appropriation that precedes all garden making. It captures the operation of symbolic power in gardening as enclosure and private paradise. On one level, the picture of Jimmy lounging in front of his pond, the centerpiece of an arrangement of mixed materials taking up what appears to be a ten-by-fifteen-foot space, captures the ephemeral photographic moment in the urban garden of a poor man of color, a squatter routinely driven from public spaces in New York. On another level, it captures the results of aesthetic plans realized: here we see design, maker, and intention all at once. Interpretation, as the sociologist Pierre Bourdieu held, requires interpretive frameworks that render symbolic power visible. One of the ways Bourdieu defined symbolic power is "that invisible power which can be exercised only with the complicity of those who do not want to know that they are subject to it or even that they themselves exercise it" (*Language and Symbolic Power* 164). Is, finally, the power of the garden laid bare by photographs of homeless gardeners nothing more or less than the inadmissible power of the social system of property and land, projected by the viewer, displaced as a pathos of dispossession?

Balmori and Morton's book includes a section devoted to "appropriated gardens," referring to gardens made by local residents on public land without legal permission. It presents appropriation as somehow separate from the making of other gardens; in other words, it accepts private property and ownership as normative. Balmori also notes "the greater difficulty in categorization . . . in defining the enormous vitality, energy, and fluidness in current garden making in inner cities" (49). Balmori and Morton's important contribution may be to confront readers with a link between the invisibility of transitory gardens and the unacknowledged social practices of appropriation in gardening.

The melancholy quality of many of the book's photographs belies the "testimony" these gardens were said to make on behalf of human needs. Outside the frame of *Transitory Gardens, Uprooted Lives,* the featured gardeners may have survived on the margins of society with somewhat less "vitality, energy, and fluidness" (49). The aesthetic criteria that the authors invite us to deploy when seeing the gardens of New York's poor interrupt, or at least temporarily suspend, the moralizing that often plagues pastoral visions of greening-up American ghettos. Yet the book's high-quality large format and consequent cost mean that its images and text are most often destined for museum and library consumption; the book was perhaps intended to bridge gaps in class, education, and background between reader-viewers and the homeless gardeners. It is not enough that we remember such places existed. *Transitory Gardens* provokes us to recognize gardening as a yearning, if not a right independent of property.

In a comparable ethnographic study of American community gardens, *A Patch of Eden* (1996), Patricia Hynes makes more space for gardeners to tell their stories. In the book, Bernadette Cozart, founder of the Greening of Harlem Coalition, describes how she brought together businesses for funding, as well as government and community groups for materiel and personnel, and drew on what Hynes calls "indigenous resources" to start and sustain gardens in Harlem. Cozart tells how the gardens, often led by female elders, connected people across what have been divisive generational, class, and racial boundaries. Cozart sought to repair previously broken linkages of knowledge that reside in older generations, particularly in ethnic communities: "Their grandmothers grew gardens in Virginia, South Carolina, Alabama, and Puerto Rico; kept fruit trees and put up food. But these kids didn't know how to use a shovel, loppers, or pruning saw!" (3). She continues: "Many young Harlem gardeners are just a few generations removed from the farms and gardens their grandparents cultivated in the Southern United States, Central America, and the Caribbean. 'But it takes only two generations to break the cycle of knowledge,' says Cozart, who introduces her prospective urban gardeners to the essentials of horticulture" (11).

Hynes describes how Cozart involves kids in recovering gardening knowledge by asking, "'Where's your grandmother from? Tell me about her garden. Did she grow berries and make you pies? What did she put up for the winter?' Stories of their grandmother's gardens and orchards spilled out, and she used these stories as touchstones for teaching the kids landscaping and gardening" (3). Hynes depicts Cozart as an anti–Robert Moses, referring to the controversial midcentury planner who reshaped New York City with massive superhighways that destroyed entire neighborhoods. The legacy of Moses looms large in the minds of city residents, as recorded in a *New York Times* interview with one garden activist: "'Robert Moses said you have to break some eggs to make an omelet,' said another resident, Carolyn Radcliffe. 'But we don't like the taste of this omelet. It's being cooked for other people to eat'" ("Crowd Storms Former Garden"). In contrast to planning projects that dispossess less affluent and often more diverse neighborhoods, Cozart imagines community gardens redressing racial and economic injustice in American ghettos.

At one point in the interview, Cozart proposes that the antiracist, antisexist, anti-imperialist work of the Greening of Harlem operates at a transnational scale. The community gardens expand notions of the "boundaries of her community" so that "Harlem stretches from ghettos here to shantytowns in Soweto. Harlem is global; it is replicated worldwide on every continent and in every race. And it's *women*. There have been many holocausts of women, but we are waking up" (32). Hynes's *Patches of Eden* reinforces the global connections made by Cozart; elsewhere, she investigates garden projects in prisons, schools, and neighborhoods across the country, including collaborations with Kenya's Wangari Maathai before her Green Belt Movement brought the first Noble Peace Prize to an environmental activist.

Hynes's global vision wielded notions of environmental justice to intervene in the economic and racial politics of the 1990s. The book, along with street-level organizing that took place in that decade, identified community gardens as counterweights to dispossession. The possibility of community gardens reconnecting generations has been documented elsewhere, including by the American Community Garden Association, which interviewed one New York City gardener who had worked continuously from a victory garden plot during the Second World War through to the Green Guerillas, spreading gardening knowledge and materials in the city (Guthrie and Levine). While hopeful, this vaster, more durable notion of community has also resulted from the catastrophic loss of many individual gardens. One in particular captured the attention of numerous observers and participants: El Jardín de la Esperanza, a garden made by Alicia Torres on Seventh Street in the Puerto Rican neighborhood of New York's East Village.

"Seed-bomb the Rubble": Losing Esperanza

Several eyewitness accounts have been published about the destruction of El Jardín de la Esperanza in February 2000. They demonstrate that environmental justice requires more than bearing witness to the dispossession of others; the residues of justice in such stories must be gathered into productive demands. Alicia Torres started El Jardín de la Esperanza, the Garden of Hope, in 1978. Four generations of her family grew vegetables, flowers, and herbs on the corner of Seventh Street near Avenue C, a block east of Tompkins Square, from which many homeless New Yorkers were repeatedly evicted in subsequent decades. The garden eventually acquired a Green Thumb lease, but the city failed to renew the lease so that the land could be sold to a private developer.

What the city described in public—and in court—as a plan to replace public lots hosting community gardens with affordable new housing was contradicted by other sources. In an interview with National Public Radio that aired on February 24, 2000, a local described the planned development as a "75 unit luxury apartment" building. On February 15, 2000, police in riot gear had surrounded the garden, arrested and hauled away thirty-one protestors, and then protected a construction crew who bulldozed the area. In Brooklyn later that same day, Attorney General Eliot Spitzer won a court injunction to stay the destruction of other gardens with Green Thumb program affiliation. Mayor Giuliani defended the city's actions, saying in a press conference also excerpted in the National Public Radio story: "We have a city with not enough affordable housing. If you live in an unrealistic world then you can say everything should be a community garden. A mayor has to live in the real world." Published accounts of the bulldozing of Esperanza quote Giuliani's words ironically, record the eviction in wrenching complexity, and represent those "residues of justice" that Wai-Chee Dimock has described in her work on the history of racism in U.S. literature and law.

Jose Torres, one of the original gardeners at Esperanza, recorded an audio diary of the garden. An edited version of Torres's recording aired on National Public Radio the week following the eviction. Torres described finding the lot in 1975 or 1976, when "there was no fence and everybody used to come in here and shoot up." Instead, the block was "full of bricks and junk and things . . . sinks, a lot of piping, boiler parts and everything." The Torres family worked "little by little" so that, in 2000, Torres could report from "the middle of Esperanza garden": "My mother's area is in this corner. This is her plot. I don't know everything that she grows in here but she grows some big stuff in here. She loves it. . . . These are the green peppers and these are the jalapenos. She's got tomatoes, lettuce, we've got a grape vine here . . . this is so beautiful that people would be passing by and we'd be cooking in here and they'd see the smoke going up . . . and you could see that they want to come in and we'd say, 'Come on in.' . . . This is beautiful, we love this."

For Jose Torres, the Garden of Hope was a beautiful, open place inviting to look at, a place where children "would hunt Easter Eggs" and gather at

Halloween. Esperanza became a ghetto pastoral landscape in miniature. It also catalyzed defiance and wide support, and Torres went on to explain how people came from hundreds of miles away to defend the garden. The courts, however, denied the group legal standing—the equivalent of choosing not to see people or hear their claims—and affirmed the city's right to sell the lots along with 112 others. The city "wasted absolutely no time," according to a reporter on the day of the eviction. Afterward, Torres recorded a few final thoughts, facing the plywood barrier in front of the razed lot. He seemed shocked, his voice notably quieter, still referring to the garden in the present tense: "Well, this is Esperanza garden. They've got plywood running across blocking the whole entire front. You can't even see in. I always thought that we were going to be there forever. . . . It's like a story, it's like a story that you can see. A rich man comes out of nowhere, a lot of money, gets rid of all the poor people. I still can't believe it . . . it's like a bomb. Right here was a garden and then 'boom' and it was gone." For Torres, the destruction of Esperanza boiled down to the ascendancy of rich men over poorer ones, a structural violence captured in the image of a bomb exploding. The attitudes of pride in the gardeners' achievements gave way to awareness of class conflict, the critical perspective of ghetto pastoral.

Andrew Light, a philosopher of environmental ethics, was present at the eviction but not arrested on that February morning. From the corner he watched the protestors being evicted. Subsequently he published what he titled an "Elegy for a Garden." In it, he reflects on what the loss of Esperanza meant in terms of environmental ethics:

> This garden was not just a patch of green on a brown landscape or a clever bit of utopian protest art. It was a schoolhouse for this particular community where elders could teach the young something about their environmental traditions, their past, and also their aspirations for the future . . . the value of this garden was unique to this locale; it was tended by these residents because it was where it was and not somewhere else. It was worth the sacrifice of defending it because it was local, rather than remote. There was no "unrealistic" desire here to create gardens everywhere, as the Mayor contended, but to maintain this one in this particular place. . . . The garden helped to make this community a site for local environmental responsibility even as it eventually came to stand for the larger environmental community's dream of a greener city.

Light enriched the significance of Esperanza as an exemplar for the "larger environmental community," but he rendered the garden more abstractly, making it almost a disembodied metaphor. On the one hand, Light drew ethical implications for a wider audience (as had the gardeners themselves), but on the other hand, the gardeners' perspective fleshed out the abstract language of moral philosophy.

While Light was observing and musing on the ethical implications of the garden, the independent-media journalist Brad Will was engaging in civil disobedience to stop the bulldozers. In an essay, he described organizing gardeners to save Esperanza using guerilla theater and sit-ins, direct-action tactics learned from Earth First! Gardeners and activists, dressed as bugs and flowers, gave testimony at city hall hearings (137). The Torres family granted permission to activists with the More Gardens! group to camp full-time in Esperanza. Will described how the garden in winter became a collective project: "We built a giant *coquí* guardian in the front . . . with room inside for three to sleep, raised up ten feet with window watchtower eyes and concrete-sealed lock-boxes. In the back . . . rose a twenty-six foot sunflower made of steel with a lock-box on top." The Garden of Hope became a stage for direct political participation. The eviction, by contrast, Will remembered as a destructive blur: "We filled the lock-downs on the fence and buried in the ground. We sang to gather strength. Dawn came quick, with the special Emergency Service Gestapo cutting open the front fence. Sudden surge of police. Yelling, scrambling, friends dragged away. Cold wet smother from the fire. Soon the taste of burning steel close to my lips, and a burn on my wrist" (138). Seen from inside the garden, the state's monopoly on the legitimate use of force looked more like brutal repression.

Will described the eviction as a catalyst for creative resistance to globalization and privatization agendas. The "cold burn" of handcuffs did not deter playful guerilla gardening by eco-anarchists during the World Bank meetings later in 2000, during which, Will said, "We inspired a Guerilla Gardening Collective to hit the streets armed with seedlings and kale seeds. These anarchists didn't come to break windows, they came to break the ground" (138). Likewise, he connected the Esperanza eviction to May Day events in 2000, when a group cleared a lot near the East River. In a gesture of continued hope, "the only thing we smashed was a piñata in the shape of a bulldozer. Inside were seeds I had saved from destroyed gardens.

They scattered on the opened ground" (138). Will's tale of destruction and a garden's rebirth by radical dissemination exemplifies how community gardens sponsored literacy about the politics of garden making, rather than serving as an escape from politics and conflict. In his words, "The story of community gardens is thousands of stories" (134). Such stories record the residues of responses to unequal force that favored property interests across postindustrial cities in North America. Like the seeds saved from Esperanza, these remnants of justice served as concentrated and potent sources for future action.

Yet the social process of environmental justice requires effective representation operating through both symbolic power of the kind registered in Will's story and socioeconomic power as analyzed by Torres. To demonstrate why both elements of representation are necessary for community gardening to do environmental justice, consider the arrangement of symbolic and socioeconomic power in a different community garden in New York, one funded by private charities: the Bronx Frontier Development Corporation or South Bronx Frontier (SBF) project.

The South Bronx gained ignominy during the 1970s as an arson-blighted moonscape bleeding population and jobs and attracting drugs and violent crime. Two community activists intervened to create a community garden on a swath of the area's real estate, bulldozing rubble and starting a large-scale composting site to remediate the soil (Jonnes 300–303). The SBF project incorporated as a nonprofit, employing many local teenagers. Although it never managed to be financially solvent, it generated tremendous symbolic power in the local and alternative media. For participants interviewed in the counterculture newspaper *Mother Earth News* in 1978, the project served as an indicator that American ghettos could regenerate communities. The historian Jordan Kleiman has also noted that community organizers sought to redeem the scene of industrial decay through eco-friendly technologies, such as the windmill that SBF installed to generate its electricity.

Yet the journalist Robert Fitch saw the community gardens of the SBF project in a different light when considered in the longer context of the city's history of politicized deindustrialization. He described SBF as green-washing the anti-industrial labor-privatization agenda of its funders. The successes of SBF depended on support from the Rockefeller

Foundation, among others. Following the money, Fitch uncovered internal memos from the foundations to support his claim that private "charities," operating at the behest of real estate developers, encouraged "community self-help" as a public relations strategy during the 1970s.

Meanwhile, during city budget negotiations, members of the finance, insurance, and real estate (FIRE) elites fought investment in public services. The real estate lobby (whose relatives often ran the charity foundations) pressured the city to grant tax abatements for their own building projects, including more upscale private housing, while fighting public funding of substantial industrial redevelopment of the area. In this reading, a large community garden such as the SBF project complemented gentrification, or the process by which the socioeconomic power of FIRE interests in the city have reshaped New York boroughs to favor the speculative building of offices and new residences, replacing industry and worker housing with upscale services. The creation of middle-class neighborhoods in the South Bronx—which were not destined for longtime (mainly black) residents able to work their way into better homes and gardens, but for newcomers—stands as one legacy of the success of gentrification. In some sites, community gardens have been used by developers to suppress broader support for residents' resistance to the privatization and deindustrialization of working-class neighborhoods.

Brad Will's partisan account of seed bombs salvaged from a destroyed garden and redeployed in a vacant lot during antiglobalization rallies stands for the radical hopes encapsulated in these fragmentary and dispersed stories. Will's account makes a stronger case for Esperanza as one of Helphand's "defiant gardens" or Hou and Lawson's "insurgent public spaces." It also underlines a kind of heroism that, as the critic William Empson proposed, was at the root of all versions of pastoral. In the context of the Lower East Side and New York's activist traditions, Will's story echoes the radical politics of ghetto pastoral that Michael Denning interpreted in the work of the city's proletarian writers. Environmental justice as redistributive power—Will's metaphor of the seed bomb—confronts its practical limits in the sphere of physical force: fences, police, bulldozers. Seed-bombing the rubble may also be a potent metaphor for the intergenerational work of environmental justice in the postindustrial context of America's deeply divided cities.

Community gardens demand and depend on different means of representation, including the revival of ghetto pastoral as a radical aesthetic project. Less radical institutional reforms can also support the long-term tenure of urban community gardens. For example, the public-private land trust model, along with guerilla gardening, has helped many New York community gardens endure a decade of rapid gentrification. These institutions do not restructure socioeconomic inequity; given these limitations, what difference for environmental justice do such community gardens make? The legal system arbitrates disputes over geographical unevenness in the city, and in its language of justice it adapts a significantly different scale of "environment" in its findings. If we shift cultural domains from home to garden, scaling up environment from family to community, a different dimension of environmental justice appears.

Environmental Justice and Reviving Hope

This chapter began with a story from Los Angeles about the South Central Farm at Forty-first and Alameda, a story that resembles in many ways the fate of Esperanza six years earlier on the opposite coast. South Central Farm had a complicated history of earlier environmental justice activism, a history linked with demographic change in a multiethnic neighborhood. In 1986 an environmental justice organization, Concerned Citizens of South Central, led by primarily African American residents, had successfully resisted the city's plan to build a trash incinerator on the site (Bullard 35; Pulido 18–19). The city acquired the land through eminent domain sale from the previous owner, the real estate developer Ralph Horowitz. In the wake of the environmental justice group's victory, the city left the site vacant for several years. In 1992, after the Los Angeles riots against police brutality, a group of Mexican American gardeners planted edible cactus on the vacant lot. The guerilla gardening prompted the city Harbor Department, officially responsible for the public land, to issue leases and sanction the community garden under the authority of a nearby food pantry. The ethnic composition of the leadership during the community actions in 1986 (led by African American residents) and 1992–2006 (led mainly by Latino residents) also represented a demographic shift in the neighborhood, from a black majority to a Latino majority, a change reflected in the city at large, as recorded in the U.S. Census data from 1970 to 2000. After peaking in 1980, the African American share of the population declined by over 15 percent.

In the mid-1990s, Libaw-Horowitz Investment Company (LHIC) approached the city to reacquire the site. The company had been granted the right of first refusal for any future sale of the land back into private hands following the earlier eminent domain proceedings. As a result of the ensuing negotiations, in 2003 the City Council approved a sale at a below-market price. South Central Farm gardeners were handed an eviction notice. They determined to dispute the eviction and initiated a series of lawsuits and appeals, through which it came to light that the city had not followed the legal requirements of public-use review in its charter for the sale of what was still public land. The private developer's subsequent attempted repossession of the site sparked a public drama. After the eviction notice of March 2003, gardeners held off removal for three years. Allies created a website, found sympathetic celebrities including Danny Glover and Joan Baez to champion their cause, and waged a public campaign against the developer through media and legal channels. When the California supreme court refused to take up their appeal, gardeners resorted to civil disobedience. Actions included round-the-clock occupation of the farm, culminating in June 2006 in the forcible removable by police of over forty activists, including seven gardeners who chained themselves to the fence.

If the city and developer ultimately triumphed through combined institutional clout and physical force, the gardeners' strategy involved appealing to the symbolic power of a counternarrative about the relationships between Latino residents as transplanted *campesinos* and the South Central Farm as a symbol of Latino political influence in the city, represented by the election of Antonio Villaraigosa as mayor, which occurred during the garden's occupation. Gardeners insisted that their right to use the site was in the legitimate public interest, protected by due process, although such claims relied less on law than on the symbolism of land-based self-reliance in American culture. When legal and cultural appeals through symbolic power failed, the activists borrowed strategies of resistance that had been tested in community gardens in New York a decade earlier. One protest banner read simply, "South Central Farm feeds families." Gardeners and their allies had first expressed the relationship of food justice between land

and community in the cactus, peach trees, and epazote planted in 1992. By 2003, they represented and disseminated this vision in handbills, rallies, digital photos, videos, and blogs. Gardening and independent media were reciprocating means to reproduce a community. Here, low-income people grew their own food in a neighborhood where bodegas stocked only packaged junk food and a few wilted vegetables. Here, an open space provided a leafy refuge for children and a gathering place for generations. In a neighborhood with scarce public resources, here seemed to be the beginning of a more just community.

While compelling, these defiant stories competed with other narratives linking people and land in South Central Los Angeles. Voices from the farm often articulated the place as representing indigenous and Chicano identities. Staking the farm on ethnic identity may have weakened support for the gardens in the surrounding community, which was diverse and historically had been predominantly African American. Despite expressing support for the farmers, Mayor Villaraigosa and local council representative Jan Brower also answered to other constituencies. Without a common mandate from a coalition of Angelinos or broader political support, local politicians were unable to effectively resist court-sanctioned sale of the land to LHIC.

The eviction struggle grew uglier as it revealed political fractures along ethnic lines, an element that Scott Hamilton Kennedy's documentary *The Garden* exposed in informal footage of garden leaders whom he unfortunately did not question further on camera. An anti-Semitic comment about Ralph Horowitz, the primary owner of LHIC, was attributed to Tezo, a central figure at the South Central Farm, and widely publicized by local media, including the *Los Angeles Times* (Hayasaki). Though the accusation may have been unfounded, as a published rumor it damaged the public perception of the gardeners and their cause.

In June 2006, bulldozers razed the South Central Farm. Eviction then joined the two main themes of community gardens in America—land as property and land as community—for a final moment when the garden ceased to exist. As in other erased urban gardens, a scene of expulsion knotted property and community together where they might otherwise diverge into distinct narratives. For the story of land as individual property—Lockean root of liberty for all peoples, regardless of ethnic background and engine of economic development—to continue, another narrative of land as collective identity, root of community and source of biophysical reproduction, has been suppressed.

In the intervening years between Esperanza and South Central Farm, thousands of new community gardens have been cultivated in North American cities and around the world. These gardens face many of the same tensions between public interest and private property, urban pastoral fantasies and versions of ghetto pastoral that compete as languages of justice. In Madison, Wisconsin, a local community garden cultivated south of the city by many Latino and white families, the Drumlin Community Gardens, came under development pressure in 2008. The rich loess soils deposited along the glacial drumlins of southern Wisconsin promised these gardeners rich harvests, barring human or climatic catastrophe. The construction of a technology center and hotel loomed as the immediate obstacle. At a rally I attended in the spring of 2008, gardeners and organizers acknowledged that the threat of destruction to the garden had become generic, with a stereotypical corporate developer ranked against minority gardeners in a less affluent neighborhood. Even in an otherwise progressive city—one often described as an island of utopianism surrounded by reality—environmental inequalities persist and continue to pass unnoticed.

By now it should be clear that the resemblance among these stories is structural rather than superficial. It underlines one final problem: to be meaningful, ghetto pastoral must provide residues of justice that generate enduring cultural traditions capable of affecting outcomes for threatened gardeners. The generic narrative must be shaped tactically to engage common political and economic structures with local factors and personalities. In the case of removal of community gardens, precedents exist for successful relocations. Several of the South Central Farm members managed to secure a new site. Because their community persisted, supported by and sponsoring a strong narrative of gardens feeding families, these gardeners transplanted their work to the new location.

Community gardeners from Los Angeles to New York, and from Madison to Milwaukee to Chicago, have generated a language of environmental justice that reconnects land and people. In twenty-first-century America, social movements that link land, food, and consumers have mushroomed. These include Community Supported Agriculture and urban farms, Slow

Food chapters and locavore dining groups—all reinvigorating political movements that Warren Belasco has traced to 1960s counterculture taking on the food industry, with wider environmental implications (19). Gardens have become places where the interests and tastes of this plural movement and consumers and producers intersect. Many such spaces have also disappeared during this time, usually for much more mundane and less dramatic reasons than police removal. Organizers move away, older garden leaders retire, younger gardeners lose interest or find themselves too busy to water tender annuals every summer day.

Contemporary stories about urban community gardens and small urban farms by novelists, journalists, scholars, and garden activists further connect the seasons of garden produce to environmental justice, as we will see in the next chapter. Reading these stories has also revealed how their cultivation has been inseparable from the practices of storytelling, and that making coherent narratives about these places sometimes precedes other demands—for economic fairness, access to good food, and livable environmental futures for everyone.

6 SEEDING NEW TERRITORIES

In the last quarter of the twentieth century, many Americans cultivated urban gardens in defiance of vicious cycles of real estate speculation, deindustrialization, and public sector divestment that perpetuated racist economic inequality. The term *environmental justice* emerged during this same period, articulated first by African American, Native American, and Latino communities with firsthand experience as the target of institutional racism across the United States. Activists identified how mainstream environmentalists framed "the environment" as an object of legal protection and spiritual veneration in ways that segregated the political interests of social groups, to the detriment of the darker-skinned, the less affluent, and the very young and very old. Engaged social scientists, religious groups, and, eventually, historians and cultural critics subsequently took up environmental justice (and the related category of food justice) as an analytical category and research agenda. By the new century, environmental justice named a widespread demand for equity in an age of widening but wickedly uneven environmental risk. By 2010, a host of long-term challenges associated with a postindustrial society combined with two decades of regressive economic policies to produce a perfect storm of poverty and hunger.

The 2000s were also marked by renewed efforts toward more pluralist and global thinking by American researchers and writers. Literary and cultural critics called for a planetary awareness of social-environmental problems, what Ursula Heise named "ecocosmopolitanism" (65), to distinguish it from localist environmentalisms. Wai Chee Dimock argued for seeing U.S. culture "through other continents," and Joni Adamson, Mei Mei Evans, Michael Bennett, Andrew Warnes, and others revealed a deeper history of environmental justice in Native American and African American literature.

Yet politically, environmentalism seemed moribund. In 2004 it was even notoriously declared "dead" by Ted Nordhaus and Michael Shellenberger for its failure as a political force capable of addressing global warming. Meanwhile, international researchers, governments, and NGOs headquartered in New York and Washington, DC, generated images of climate apocalypse and resource wars. These competed with utopian descriptions of smart green cities, with towers of glass linked by hanging gardens and planted promenades, funded by a new era of unlimited growth. Bullish and bright green fantasies dueled with doom and gloom—especially in Hollywood—right up to the brink of a global credit crisis in 2008 that sent unemployment, home foreclosures, and poverty rates in the United States to levels not seen since the Great Depression. In the context of so much social wreckage and ecological uncertainty, American gardens once again took on an overdetermined set of salutary and symbolic political meanings. Like earlier economic contractions in the 1930s and 1970s, the 2008 bust launched a generation of homesteaders, but this time many preferred a different term for their work, *urban agriculture*. A signature quality of what Michael Denning described as the interplay in ghetto pastoral of "degradation and elevation, the grotesque and the simple" (251) describes well the dark mood occasionally leavened with ironic laughter and wild hopes that characterizes U.S. garden writing in the new century.

This chapter examines two garden-based memoirs from the first decade of the twenty-first century: Jamaica Kincaid's *My Garden (Book):* (2001) and Novella Carpenter's *Farm City: The Education of an Urban Farmer* (2011). Both writers belong to the post–civil rights generation and came of age in a postindustrial society, with its attendant instability, migrations, and postmodern aesthetic. Yet through family memories, their narratives stretch back to the great postwar economic expansion and progressive political emancipatory movements of the 1960s and 1970s. In their shared interest in a public and ethical dimension of gardening, both authors bring forward the tradition of garden writing that represents an uneven but evolving discourse of environmental justice in America.

Jamaica Kincaid speaks from the geographical region that is familiar to us from Katharine White, Scott and Helen Nearing, and Michael Pollan's earlier writings based in rural but declining agricultural landscapes in the Northeast. Her later vision is markedly more ambivalent. In framing contemporary American gardens through postcolonial ecologies, she emphasizes the complicated politics and expansive desires involved in immigrating to the United States and acquiring land in postagrarian districts. Kincaid finds gardening in America unavoidably a postcolonial activity, dependent on global horticultural reach, legacies of ecological imperialism and territorial conquest, and complicated histories of painful economic mobility.

Carpenter brings a countercultural, wry sensibility to farming in the city. Allison Carruth argues that Carpenter's book is a "locavore memoir" that exemplifies how contemporary food writing links global networks of knowledge and concern, what she aptly names "the precarious, open-source future of urban food production" even when menus and politics are militantly local (*Global Appetites* 161). This open-source future often turns to urban history and residual country ways, and in fact Carpenter's work relies heavily on the genealogy of garden writing and calls for a rebirth of organic cities. These cities respond to economic and environmental inequalities by prescribing more public access to gardening and, ultimately, urban farming. But like Kincaid, Carpenter's garden narrative struggles to maintain a comforting arc of personal development and something like middle-class security. These memoirs cover political strategies for cultivating gardens. Kincaid forces readers to maintain a split consciousness of imperial legacies and class divisions, while Carpenter's community garden comes to depend on ties of solidarity and a broad base of community support in a new "Gilded Age" of vicious inequalities.

The Postcolonial Ecology of the "Conquering Class": Jamaica Kincaid's American Garden

My Garden (Book): epitomizes the multiscalar consciousness that Ursula Heise called ecocosmopolitanism, a worldly awareness of the global forces that interpenetrate every local ecological transformation and social act (65). In *Sense of Place, Sense of Planet,* Heise at one point asks rhetorically: "What if work and hearts are not confined to continents but sustain ties to several?" (44). Kincaid's book offers a cultivation narrative that shows how American gardens tie their makers elsewhere and link us to historical injustices behind the appropriation and transplantation of entire botanical worlds between Europe and North America. She suggests a role that garden writing might play in a twenty-first century marked by increased division between life outcomes in the Global South versus the Global North and expanded consciousness of the transnational context of environment and culture.

Kincaid probes the ambivalence of gardening in the United States with a split consciousness of its joys and costs. Born on the island of Antigua, Kincaid moved to New York City as a young woman and began her writing career at the *New Yorker*. Once established as a successful novelist in the 1990s, she began plotting a large garden around her Vermont home. *My Garden (Book):* works at the crossroads of declensionist and progressive metanarratives of American nature.[1] In contrast with the progressive arc of U.S. social history—including celebrations of the rise of 1970s environmentalism—the narrative arc of much environmental history remains declensionist. In this metanarrative, environmental quality has declined across the entire landscape as Americans extinguished or decimated species after species (passenger pigeon, bison, wolf, monarch butterfly), paved much of the continent, and became increasingly alienated from nature.

In his history of the U.S. environmental movement, Philip Shabecoff has emphasized the ways in which the complicated mingling of these metanarratives has meant that environmentalism often defies the left/right binary of American politics. Shabecoff observes that environmentalism is both progressive (human society can be improved; we can mitigate climate change) and conservative, in the sense that "the desire to cultivate a simpler lifestyle and preserve the landscape and the quality of life, are distinctly conservative goals" (238–39). Shabecoff's reading complicates a standard account of environmentalism as inherently an extension of liberalism, a granting of rights to progressively more entities from a minority of people to all people and then to animals, plants, and habitats. It suggests the necessity of Kincaid's postcolonial awareness of American nature as an environmental-historical artifact shaped by supremacist and imperial legacies of oppression. Nowhere are the ramifications of past environmental injustice and colonialism more evident than in gardening.

In *My Garden (Book):* Kincaid explores how the making of her American garden falls between these two metanarratives. For the author, gardening is always a hopeful gesture and an "exercise in memory," often the painful remembering of imperialist trade routes of enslaved humans, seeds, and ideologies (8). For, as she puts it, "Memory is a gardener's real palette; memory as it summons up the past, memory as it shapes the present, memory as it dictates the future" (219). In exploring her own garden making in the sweep of colonial history in the Antilles, Americas, and beyond, Kincaid works through the situational irony of her expansive desires as a gardener and the botanical consequences of imperial conquests. Her book makes a global garden tour, circulating from her present garden in Vermont to childhood gardens in Antigua (including her family's home grounds and the British botanical collection) to more recent excursions through England and the Himalayan region of China.

Kincaid often interrupts the transnational movement of her global garden tour by reflecting in the first person on the psychological frictions of gardening as a postcolonial immigrant and naturalized American. Surveying her Vermont garden, she realizes "with a certain amount of bitterness" that its pleasures, derived from the freedom and knowledge she has as its creator and from the physical work that makes "her body a cauldron of smells pleasing to her," would make her a "picture of shame" in the place she is from (121). In becoming a gardener, she feels she has crossed a threshold: "I thought how I had crossed a line; but at whose expense? I cannot begin to look, because what if it is someone I know? I have joined the conquering class: who else could afford this garden—a garden in which I grow things that it would be much cheaper to buy at the store? My feet are (so to speak) in two worlds, I was thinking as I looked farther into the garden.... To me, the world is cracked, unwhole, not pure, accidental; and the idea of moments of joy for no reason is very strange" (123–24).

One of the most compelling techniques on display here is the narrator's repeated self-questioning about what it means to become a gardener in the United States: what one loses in moving from Antigua to Vermont, from being one of the conquered to one of the "conquering class." These are questions of environmental justice as much as they are questions of class, race, and nationality that arise from the contingency of her garden's cultivation. For Kincaid, and by extension her readers, they also disturb and estrange gardening's joys, including the capacity of botanical beauty to absorb our full attention, by vitiating their innocence. "And what is the relationship between gardening and conquest?" she asks pointedly (117). Kincaid views English gardening in the New World as a material and ideological tool of political conquest.[2] In remembering childhood Sundays spent at the British royal botanical garden in Antigua, Kincaid unpacks its association of aesthetic pleasure, botanical knowledge, and conquest. She remembers seeing "plants from various parts of the then British Empire, places that had the same climate," but "none of the plants were native to Antigua." She continues: "The botanical garden reinforced for me how powerful were the people who had conquered me; they could bring to me the botany of the world they owned" (120).

As she frames her twenty-first-century Vermont garden in this international context of imperial gardening, she evokes island landscapes layered with histories of conquest, appropriation, and desire. As Elizabeth Deloughrey points out in "Island Ecologies and Caribbean Literatures," Kincaid deliberately disrupts a unified chronology linking pre- and postcolonial botany in Antigua in order to draw attention to the temporal and spatial work of imperial power (304). Kincaid also presents an American garden animated by historical conflicts and a gardener with a sensitivity to the class privilege and racial politics of possessing and cultivating land. Her book represents a darker, more ironic version of garden writing than Pollan's utopian vision in *Second Nature* a decade earlier, wherein cultural-botanical diversity in the "blooming archive" symbolically resolved ethnic and class divisions. Kincaid's more skeptical view of the construction of nature via acts of botanical naming, possession, and racialized conquest reinforces a larger environmental justice argument made by a host of contemporary environmental historians and literary critics. Women, African Americans, Chicanos, Native Americans, queers, and other subordinated social groups were for most of the twentieth century systematically written out of dominant concepts of nature epitomized by the heroic masculine wilderness ideal.[3]

Kincaid gives readers reason to be skeptical about how we might, through reflection and questioning alone, arrive at an untroubled cosmopolitanism through gardening. She shows how her own desires as a gardener have proved unstable and elusive. In a phrase that is characteristic of the tangled, vining style of *My Garden (Book)*: Kincaid writes of gardeners: "They only love, and they only love in the moment; when the moment has passed, they love the memory of the moment, they love the memory of that particular plant or that particular bloom, but the plant or bloom itself they have moved on from, they have left it behind for something else, something new, especially something from far away, and from so far away, a place that they will never live (occupy, cultivate; the Himalayas, just for one example) (218–19)." Her choice of the Himalayas is exemplary and personal. At the end of her book, Kincaid describes plant hunting in southwest Yunnan, China, and the shock of finding mountain forests of rhododendrons "with felted leaves," a variety she knows from a specialist nursery in New York and that in her Vermont garden had "seemed so exotic (coming from far away, unfamiliar, rare, hard to come by)" (224). Crossing borders, Kincaid finds herself in a landscape that reminds her of "Eden, or something like it, only this time turned inside out," for "this time (in China) the garden was in a state of banishment; I was in the wild, the garden had become the wild and I was in it (even though all the time I was really in China)" (226). In a progress interrupted repeatedly by parentheses, Kincaid reaches for a transnational notion of gardening that does not recapitulate the blinded perspective of a conquering class, but instead challenges limited knowledge and expansive desires.

Later, Kincaid writes, "I had (have) come to see that a garden, to make a garden, is partly an attempt to . . . bring in from the wild as many things as can be appreciated, as many things as it is possible for a gardener to give meaning to, as many things as it is possible for the gardener to understand" (226). To make a garden in the United States in the twenty-first century means cultivating worldly appreciation while understanding how the desire to cross borders in search of the wild, the "unfamiliar, rare, hard to come by," has been (and is, to use Kincaid's emphatic shift in tense) linked historically with conquest.

Chinese fringed rhododendrons end up for sale in a Vermont nursery at the end of a trade route opened generations earlier by British military force. Such a displacement of naturalized, gardened landscapes defines what the literary critic Jill Didur, in a reading of Kiran Desai's novels, calls the work of "literary postcolonial ecology—where decentered notions of belonging and localized modes of existence might thrive" (59). Kincaid's postcolonial ecological perspective of the planetary territorial projects of Anglo-American gardening lights the pathways by which historical conflicts result in the particular ecology of her contemporary suburban American garden. Such garden writing, as a valuable critique of the conquering class, helps us interpret more widely marketed narratives of cultivation that link manicured nature and class mobility in suburban America.

Gardening as Economic Justice: Reviving the Organic City

Kincaid brings a postcolonial edge to an autobiographical, reflective mode of suburban garden writing that was also pursued by Lawrence, White, and Pollan. Other twenty-first-century gardeners and garden writers, influenced by the community gardening movement, have become the critics, informal public historians, and utopian revolutionary designers of U.S. cities. Rather than meditating on the historical routes of botanical colonialism, a small but vocal number of garden writers, theorists, and filmmakers have engaged with the burgeoning discourse of ghetto pastoral that I described in chapter 5. These individuals joined the community garden movement and added to a growing repertoire of activities aimed at building more equitable cities. Writers and activists created seed-saving cooperatives, expanded vacant-lot gardening, founded urban homesteads, and incorporated animals into urban gardens.[4] Honeybees, laying hens, and even rabbits and pigs returned to neighborhoods from Milwaukee to Brooklyn, Oakland to Atlanta, after a century of exile following what the environmental historian Ted Steinberg, in his 2002 book *Down to Earth*, describes as the "death of the organic city" (157). Perhaps the most widely read of this new brand of urban garden writer is Novella Carpenter, whose *Farm City* became a national best seller in 2011. The book's success is certainly due in part to its feisty tone as well as for the timeliness of its explicit call for reviving the organic city and its implicit demand for economic justice.

According to Steinberg, the term organic city describes a mid- to late nineteenth-century system in which the working poor (and especially doubly marginalized groups: poor children, African Americans in northern cities, and new immigrants) recouped value and livelihoods by recycling energy and materials within sprawling, often filthy cities. Steinberg's use of the term organic designates relations of exchange between human and nonhuman residents as they produce and reincorporate organic matter as food, waste, and raw material. The phrase also gestures toward a quality of interdependence and social holism among poorer city dwellers prior to the sanitary reforms of the late nineteenth century.[5] In the back lots and narrow alleys between creaky antiquated tenements, "necessary tubmen," predominantly African Americans in northern cities, emptied privies and sold human waste to truck farmers at the metropolitan edge (164), while "swill children" collected household scraps to feed backyard pigs (170). Other residents of mid- to late nineteenth-century cities participated in a brisk trade in rags, bones, and scrap metal—those fragments of a vanished material culture that inspired gritty imagery in literature from Dickens to Yeats.

The ecology of these organic cities involved tremendous risks for urbanites, specifically in the lethal circulations of malaria, cholera, and other water- and waste-borne infectious diseases, which put not a few swill children in the pauper's corner of municipal cemeteries. Steinberg is careful to point out the virtues of cleaning up the organic city while acknowledging what was lost in exchange for "the great cleanup" (171), including the lost soil fertility for truck farms and home gardens. The mainly upper-middle-class reformers who led the public health movement cast out backyard pigs and mandated changes to building codes that eliminated economic niches based on organic waste recycling. The sanitized and industrialized city effectively exiled food gardening and urban agriculture for several generations.

Yet industrial cities remained subject to cycles of energy production and decay—ecological processes that are simultaneously political-economic conditions. Without the continued capital investments in downtown infrastructure characteristic of expanding urban areas from the Progressive Era to the mid-twentieth century, cities transformed once again. In the 1960s and 1970s, inner cities from Newark to Oakland did not immediately revert to organic cities with make-do ecologies of subsistence; rather, they became a texture of lower-cost convenience stores, run-down housing, abandoned factories and warehouses, and empty lots. Meanwhile, most new residential building and demographic growth shifted to the sprawling suburbs. James Howard Kunstler dramatically labeled this patchwork landscape of megastores and malls "the geography of nowhere," the title of his 1993 book. Urban garden writers and activists reacted against the presumed nonidentity of such places by seeking to rebuild inner-city neighborhoods.

In the 2000s, popular discussion of the place of gardens in revitalizing cities took place within a context of urban redevelopment schemes and tensions around gentrification, or the economic and cultural transformation of older urban neighborhoods through an influx of capital and wealthier inhabitants. As with the Lower East Side in New York during the 1990s, gardens remain within the urban fabric of many gentrifying neighborhoods, either through local organizations with roots in self-reliance community politics or through corporate-funded initiatives. Gardens and urban farms have also been cultivated in cities harder hit by economic and demographic decline, such as New Orleans and Detroit. Advocates for backyard bees, chickens, community gardens, and urban farms supporting goats and pigs in downtown ZIP codes often write and speak of regaining contact with organic processes, earthy fertility, and, of course, neighbors. This seeming rebirth of the organic city has joined in the processes of gentrification and suburbanization. In grappling with the ambivalent status of urban agriculture, Novella Carpenter contributes to ecological and social critiques of postindustrial American cities, notably by emphasizing the creative potential of communities rather than the yawning deficits of environmental inequalities.

Carpenter's *Farm City: The Education of an Urban Farmer* presents itself as a critical rewriting of Michael Pollan's *Second Nature: A Gardener's Education*, a direct response given that Carpenter studied journalism with Pollan at Berkeley while working several part-time jobs, gardening, and writing the *Ghost Town Farm* blog that fed into the book project. Carpenter interrogates her parents' homesteading in Idaho in the 1970s and describes her garden in Oakland's Ghost Town neighborhood as a community hub, an avant-garde creation, and an alternative to the chic new urbanist commercial development creeping into the town's edges. Her garden and farm are planted on the dark side of the postindustrial city, in the shadow of the bright sites of consumption for affluent nonresidents able to externalize production impacts. More precisely, the book takes place in a district of Oakland characterized by the geographer Nathan McClintock as a social geography of inequality, with poorer residents of color living among food deserts in the flatlands and wealthier, whiter residents residing in the hills (90–91). McClintock uses the USDA Economic Research Service[6] standard for defining a food desert as an area "with limited access to affordable and nutritious food, particularly such an area composed of predominately lower income neighborhoods and communities" (qtd. in 89). Oakland presents itself to Carpenter as cluttered with its own miraculous fragments of cultivated nature, from the author's squat garden to fruit trees in abandoned yards. "I have a farm on a dead-end street in the ghetto," she declares (9), describing it as "the third world embedded in the first" (12). Her perspective on the socioeconomic conditions of a deindustrialized ghetto involves transplanting to an urban space ideas and practices of self-reliance, domestic economies of use, and the dissenting politics of her homesteading parents.

Carpenter describes her decision to move to Oakland and start an urban farm as inspired by her parents' Idaho homestead: "I had just reached the age of my parents when they started farming and I felt drawn back to the ranch," a "house my parents had built with their own hands: a rough-hewn cabin covered with cedar shingles and a tar-paper roof," which she finds has burned down, the "apple trees had gone feral," years after her parents stopped working the land (144). Yet there is more to Carpenter's decision to convert an empty adjacent lot into a poultry run and pigsty than pure nostalgia for a childhood home. She contextualizes nostalgia with an intergenerational perspective on the potential of liberating political and social movements by describing her mother's disenchantment: "Along with most of the other back-to-the-landers, my mom had realized that the remaking of our entire American society might not be possible in her lifetime. . . . The back-to-the-land movement's failure, as inevitable as the collapse of every other utopia, became a buffet of schadenfreude at which even I had occasionally feasted" (116). Carpenter's ambivalence toward the feminization of much homesteading work, including gardening, echoes the sentiments reported by many female participants in the back-to-the-land movement of her mother's generation (Unger 172). Yet the edginess of her writing style—she swears and dishes slang generously on the page—also marks a departure from the decorous tone of earlier generations.

Such intergenerational connections and tensions underline the complexity of reintroducing gardens and food animals into American cities;

rather than a simple nostalgic return, this is a radical environmental aesthetic. *Farm City* also draws from an earlier generation of feminist environmental and avant-garde artists, such as Martha Rosler and Mierte Ukeles. These and other artists systematically estranged domestic space and labor through public performances of household chores. The praxis of art and politics and a call for institutional change were central elements of Ukeles's antimanifesto for "Earth Art" in 1970; her contemporary and New York neighbor, the artist Liz Christy, realized a different form of earth art in the creation of the city's first recognized community garden. Such works are now recognized as the vanguard of environmental art for the way in which they dissolved presumed boundaries between art and life, emphasized radical rupture with the past, and performed in dramatic and provocative fashion everyday work of socioecological reproduction from perspectives that instantly estranged it. *Farm City* also reveals social reproduction as an ecological process. Carpenter incorporates an aestheticized, avant-garde attention to the materiality of housekeeping and domestic life, its remnants of food and landscapes, into a seasonal, autobiographical garden book.

If cleaning and mending are part of Carpenter's memory of her mother's homesteading work, gardening presents itself as a messier, more creative form of toil. Inviting neighboring species into the house becomes part of the urban farmer's education. Most of *Farm City* takes place in an apartment increasingly turned inside out into the Ghost Town neighborhood, acting as an incubator and hive for feeding neighbors and drawing people together. On a visit to another neighbor who keeps chickens, she notices how farming in the city seems to require this creative mess: "I snuck a peek into the woman's house while Willow wrote her check. Yep, it was as disorganized as mine. Egg cartons piled up on her counters; feathers wafted to the floor. Now that I was part of the farming club, I had come to the conclusion that farming isn't without its downsides. Like my filthy house, for instance. Between my various real jobs and animal husbandry, there was just no time for cleaning" (209). Carpenter takes this messiness as "a sign of a busy, full life." It is this positive appraisal of disorderly fecundity, rather than the strict disciplining of domestic hygiene, that characterizes a revival of the organic city.

Yet ultimately, Carpenter explains her growing garden and animal husbandry in economic rather than aesthetic or moralistic terms. She describes depending on multiple low-paying jobs in the service industry and getting into the backyard chicken scene in order to eat well on a budget. Here Carpenter's *Farm City* in an Oakland neighborhood marks a qualitative shift from the earlier organic city. She faces similar economic necessities (working for low wages with scant social support) and finds a comparable solution (eventually feeding two pigs on Dumpster refuse). Yet Carpenter's quest for healthy, high-quality food leads her to critique the industrial food system and racist socioeconomic inequalities that have left the children in her neighborhood with more intimate knowledge of liquor stores and fast food than the carrots and melons she grows in her garden.

Moreover, Carpenter's Duroc pigs and heritage-breed turkeys have names, an unremarkable fact for the 4-H kids at the livestock auction where she buys her hogs, but one that represents a self-conscious revaluing of residual culture. In telling the life stories of individual animals, starting with meat birds such as Harold the "postmodern turkey" (91), Carpenter defines an ironic, post-Romantic environmental awareness.

She also presents a postindustrial, postmodern rewriting of earlier ideas in the genealogy of U.S. garden writing, emphasizing economic solidarity with her neighbors and sharing food and resources. Carpenter ends *Farm City* with an affirmation of solidarity: "We were just trying to survive" (184). She comes to identify with her neighbors and connects with several members of the original Black Panthers and other urban agriculturalists who create a local market for fresh greens at affordable prices. Through Dumpster diving at a new high-end restaurant on the gentrifying edge of her community, she meets the chef and maker of artisanal salami, who teaches her how to prepare meals and preserve every part of a pig.

The economic tenuousness of this network does not detract from its meaning. If anything, the interdependence of Carpenter's garden and animal husbandry with her neighbors defines an urban farm, just as once they may have characterized organic cities. "So when I say that I'm an urban 'farmer,' I'm depending on other urban farmers, too," she writes, adding that "it's only with them that our backyards and squatted gardens add up to something significant" (269). The specific pressures on her squat garden involve familiar forces of gentrification, as when her landlord puts the lot up for sale and real estate brokers seek to market it for condo development.

Commercial zoning of Ghost Town, itself a legacy of its more affluent industrial past, enables new businesses including a Whole Foods store to move in, which remaps the area as "a neighborhood in transition" (264). As a best-selling book, *Farm City* participates in this rebranding of Oakland as edgy with hipster chic and artsy roughness even while expressing a worldly dissatisfaction with what Carpenter earlier calls the "paradox of development." She observes that "as long as our neighborhood stayed messed up, I could have my squat garden and my menagerie" (98). Yet on autobiographical and wider political terms, Carpenter's vision of a farm city stands for a fairer distribution of nature in the form of healthy food for diverse but relatively poor city dwellers. Just so, urban gardening movements have sought environmental justice through substituting alternative systems of value. They have specifically emphasized the social meaning of good food and the substance of knowing one's neighbors instead of the system of economic capital linked with real estate that underwrites the paradox of gentrification.

Carpenter ends her autobiography with a snapshot of the Ghost Town Farm landscape operating on a prophetic timeline discontinuous with property-value-boosting gentrification. She emphasizes the subversive potential of slow, incremental work, of planting fruit trees and building the soil. Although the penultimate chapter sounds a resigned, even fatalistic note, the final chapter ends with a resistant message: "My farm will eventually be bulldozed and condos will be built.... And who knows, maybe a few neighborhood kids like Dante will pass by the units and tell someone who doesn't care, 'There used to be a farm here.' Maybe the peach trees planted in the parking strip will remain, and a hungry urban forager will cherish the ripe peaches someday. The soil here will be uncommonly abundant, and maybe someday a strange-looking vegetable will sprout here again, when the moment is ripe" (267).

The moment was ripe in the United States for a story like *Farm City* when the book hit shelves in 2010. Its themes of making do with less, finding meaning in solidarity with one's neighbors, and coming to terms with deep economic uncertainty resonated with many Americans in the trough of the Great Recession and made the book a best seller. This modest economic success in turn enabled Carpenter to rewrite the ending. She later reported on her blog that with money from the book sales, she and her partner were able to purchase the garden lot from its owner.

Does Carpenter's participation in the paradox of gentrification invalidate claims that her garden is a model for environmental justice in the United States? Is there an alternative way to save the substance of a public garden in postindustrial America without commodifying it, to save it without selling it? The real-life epilogue to *Farm City* raises such questions, and given the persuasiveness and well-meaning efforts of many urban farmers, it is hard to answer them unequivocally by staying within the boundaries of Carpenter's narrative or, for that matter, within the space of gentrifying postindustrial cities.

Bucolic Futurism and Growing Power

One of the national leaders of urban agriculture, Will Allen of Growing Power, Inc. in Milwaukee, provides a useful and sometimes troubling thinking-through of these questions of development and justice in his own 2013 autobiography, titled simply *The Good Food Revolution* (coauthored with Charles Wilson). He has done so largely by demonstrating the necessity of developing more sophisticated production systems than those of small food gardens. By 2012, nearly half the African American children living in Milwaukee struggled with poverty and food insecurity. During the worst years of the recession after 2008, roughly three-quarters of the city's black men were unemployed. By any measure, economic conditions had worsened over the last generation and racial segregation had held at a constantly embarrassing level—in the top ten of American cities. Formerly middle-class and predominantly African American neighborhoods in West Milwaukee showed few tell-tale signs of gentrification. Nor had developers advanced on the "transition" neighborhood of Riverwest with much enthusiasm, though a mix of urban pioneers (artists and students) made it a vibrant place to live.

Meanwhile, persistent toxicity in the form of high lead levels in urban soil, PCBs and dioxins in river sediment, and a load of agricultural runoff along with E. coli bacteria that surge with summer floods and in sewerage overflows created an urban ecology of greater risk for downtown residents. Yet by 2000 many of the heavy industries responsible for this poisonous legacy (leather tanning, engine and engine-parts plants) had closed down or shipped work overseas.

To build a working farm within Milwaukee that could produce affordable healthy food, Allen found it necessary to partner with local universities, collaborate with charitable and religious organizations, and form a nonprofit organization. Today, an army of volunteer labor supplements the production of red worm castings, compost, organic greens, and fish produced by aquaculture systems. And as Alfonso Morales has documented in "Growing Food *and* Justice," Growing Power has catalyzed a coalition of regional affiliates (149). On the strength of decades of creative work and a much-publicized MacArthur Foundation award, Allen has made Growing Power a national hub for urban agriculture. Potent political symbolism resides in his biography as an African American son of sharecroppers in South Carolina turned professional basketball player and successful businessman, who, in midlife, returned to farming and sought to strengthen a struggling community from the soil up. His autobiography presents a more ambiguous narrative. Recalling his work in the late 1990s for a state corrections program teaching juvenile offenders horticulture in order to raise money for his nonprofit, he writes, "I knew it was naïve to think that gardening could turn around the lives of young men whose circumstances were so difficult" (127). And yet in the next decade, Allen continued to build Growing Power into a creative center, seeding dozens of new community gardens and feeding hundreds of Milwaukee residents every week from his farm.

Today the headquarters of Growing Power transforms weekly loads of food waste from Milwaukee brewers, grocers, and coffeehouses into hills of compost. The original greenhouses have gone vertical, packing more food into hanging baskets and tiered aquaponic systems through infusions of worm- and waste-fed fertility. Chickens, turkeys, and goats thrive on the premises—their pens mucked out by local college and high school students wearing knee boots. By 2012 a more ambitious vertical farm was in the works—utopian blueprints worthy of a *Whole Earth* catalog from 1969 have already been drafted in a style of bucolic futurism. Thousands of people have attended trainings, and over two thousand visitors packed into the Wisconsin State Fair Park for a conference on urban agriculture. For this aspiring minority of U.S. citizens, the promise of vertical farms built in Milwaukee became more alluring than any hanging gardens of Babylon.

Environmental justice requires a fundamental redistribution of power and money among people in the postindustrial West, exploited Global South, and internal Southlands. Ecological justice—rather than restoration only—demands the redistribution of energy and material with nonhuman creatures on this artificially warmed planet. These tasks are simply too big for a single garden and more complex than a single garden-centered ethic. Yet in the first decade of the twenty-first century, cultivating this vision of gardens as potential hubs of revolutionary change—not merely as producers of food but as models of alternative political economy, seedbeds for organic cities, sponsors of bucolic or just futures—resonated with Americans not despite but because of increasing economic disparities.

Epilogue
GARDEN WRITING AND A PHENOLOGY OF SURVIVAL

From the pages of this book it should be clear that cultivating gardens in the United States over the past century did not automatically make the country's society more just or its farms, cities, and suburbs more ecologically sound. Yet a significant minority of garden writers offered social and environmental critiques that anticipated the concerns of environmental justice. Twentieth-century garden writers wrote with uneven but growing attention to environmental inequalities and toxic exposures, problems of food and land access. The writers considered here have cultivated gardens in print as engagements with the literal and figurative political landscape, turning garden writing into fertile conceptual ground for thinking through the conflicts and compromises involved in making culture and transforming environments.

Today as in the past, garden work is divided according to gender, race, and class, with important regional variations, even while the past two decades have witnessed a renaissance of writing about ideas of environmental justice and common rights to land and food. Rather than seek to construct a unified genre, I have surveyed gardens cultivated for food, flowers, and nontangible valuables and presented examples of the strongest articulations of ideas. In asking how these texts have spoken to one another I have compared different practices of cultivation, seeing continuity as well as social differences between spadework, plant selection, and literary techniques.

Reading cultivation as a spectrum from the edible to the literary follows from the sociologist Pierre Bourdieu's invitation to analyze the social construction of tastes by reconnecting "the elaborate taste for the most refined objects . . . with the elementary taste for the flavors of food" (*Distinction* 1). For example, Katharine White's garden and essays shuttle between city and country, deploying literary taste for rural New England in which gardening is an aesthetic category to justify a paradoxically urbane country life during decades in which suburbanization was deplored as mass culture. From the opposite view of the abandoned urban core, community gardens in the 1970s provoked rival claims to public land and the public interest; a generation later, they became a gathering place for dissenting voices. Instead of hearing contradiction in these varied senses of the term *cultivation*, an analysis of garden writing has the potential to refine our appreciation for necessary uses of land while deflating the pretensions of literary cultivation and alerting us to the dangers in mystifying uses of garden metaphors.

We have seen how unequal access to means of cultivation has constrained and shaped representations of gardens from Maine to Milwaukee, North Carolina to Los Angeles. Patterns of environmental inequality emerged in the relation of garden writing to culture, history, and environment in America. Many of the garden writers in this book imagined cities as the opposite to what their gardening represented, whether they offered intellectual dissent from mainstream suburban culture or a rejection of capitalism's environmental and social destructiveness. For much of the past century, garden writing was attached to exurban flight and rural landscape in a context of urbanization and industrialization. These rural gardens provided an aesthetic and political ideal that fell short of the demand for greater distributional or compensatory justice for millions of working people, including African American, Latino, and immigrant farmworkers whose stories are only now being told. The present generation finds urban gardeners by necessity reinvigorating a revolutionary political discourse, reviving ghetto pastoral, reimagining organic cities, and demanding environmental justice for city residents. They reoccupy ground once staked out for public gardens on behalf of democratic culture by nineteenth-century reformers—a culture promised but never fully realized. Moreover, although distinct social groups may disproportionately produce certain forms of writing, the garden writing gathered here reveals much dialogue across class, race, and gender rather than a formulaic genre. Simply put, a more varied and exciting body of writing on gardens has appeared since 1980 rather than what Katharine White imagined as a unified national genre.

The contested making of urban gardens has meant that community gardeners, their allies, and critics have sought to aggregate disparate stories. It is perhaps surprising that garden writing contradicts what the critic Lawrence Buell refers to as the tendency of place-based writing "to interest itself especially in bounded areas of small size" (*The Future* 77). Garden writers in the United States increasingly foreground the global routes of plant materials and tastes that have gone into the making of their garden's boundaries. Several critique the political significance of gardening by connecting the small-scale place making of their gardens with the large-scale imagination of the region (as in Lawrence), the nation (White, Michael Pollan), and the globe (Jamaica Kincaid and, more recently, Emma Marris).

Community gardeners and urban farmers have also linked their small-scale plots to global justice movements, participating in a major shift away from mainstream institutional environmental politics toward horizontal cooperation. Like the best environmental writing and criticism, such garden writing "offer[s] the promise of correcting against [localism] by refocusing attention on place at the level of either the region or the transnation" (Buell 82). The emergence of collective forms of garden writing, increasingly transitory and deterritorialized as they proliferate through social media, requires further critical attention.

In scaling up narratives from an individual garden, however, have these stories lost some of their affective power? The most memorable urban garden stories—for example, the accounts of the Jardín de la Esperanza on the Lower East Side or Diana Balmori's portrait of Jimmy's squatter garden—derive power by individuating a struggle against structural violence for means of cultivation. As we have seen in Michael Pollan's work, focusing on an individual gardener, even while echoing the voices of many gardeners and many gardens, can be an effective way to engage environmental ethics. The first-person narrative form, however, carries the liabilities as well as the strengths of the individual perspective. These include a writer's blind spots. In Pollan's case, the "idea of the garden" as a model for decision making about how to shape landscapes assumes a degree of affluence and privilege, even as the ironic awareness of his narrative ensures that his ideal gardener breaks with the ideological, metaphorical language satirized in the political garden novel par excellence: Jerzy Kosinski's *Being There*.

The mixture of curiosity, complicity, and irony of Jamaica Kincaid's perspective in *My Garden (Book):* also tempers over-hopeful notions about gardening as a cultural and environmental panacea. Her millennial caution seems warranted given that enthusiasm for gardening exploded in the most recent cycle of a slow-motion crisis of economy and environment. In the first decade of the twenty-first century, gardens once again appeared at the forefront of politics in America. First Lady Michelle Obama's creation of an organic vegetable garden on the White House South Lawn in March 2009 provoked a public letter of protest from advocates of chemical "crop protection" (the industry's euphemism for pesticides)—and praise from environmental justice activists, slow food enthusiasts, and community garden organizers.[1] In a letter from the Mid America CropLife Association (MACA) addressed pointedly to "Mrs. Barack Obama," MACA's executive director "respectfully encourage[d] [her] to recognize the role conventional agriculture plays in the U.S. in feeding the ever-increasing population, contributing to the U.S. economy and providing a safe and economical food supply" (qtd. in Richardson). Despite representing the interests of a powerful trade lobby, the letter sounded a populist note in cautioning against not just the organic garden but the first lady tending any garden at all as a model for responsible environmental citizenship. "The time needed to tend a garden is not there for the majority of our citizens," warned the industry's boosters, "certainly not a garden of sufficient productivity to supply much of a family's year-round food needs." A chorus of recent books and articles has attacked the cynical motive behind this claim. Does the status quo (industrial agriculture leaking toxins through groundwater, soil, and air) represent a necessary evil? Or might it be possible, as Will Allen suggests at the end of his memoir, for U.S. citizens to "pick up a shovel" (249) and join the "Good Food Revolution," gardening to sustain healthy lives and communities?

This new wave of overtly political garden writing targets a host of problems: slowing global climate change by decreasing the carbon footprint of food distribution; cultivating plots in urban food deserts to equalize access to nutritious foods; slimming children's waists by having them grow a healthier lunch from school gardens. In this return to gardening is a call to renew civic activism in U.S. culture and soften its impact on the environment. Sustaining a habitable world in the next century may indeed require horticultural practices taught by a new generation of garden manuals: composting, mulching, companion planting, seed saving, and community organizing.[2]

Politicizing garden cultivation presents dangers as well. Those who have created their own ideal of cultivation through overuse of garden metaphors have been blinded to the unjust costs of transforming landscapes already inhabited by less powerful groups. The British vision of the New World as a garden justified that nation's imperial conquests; the Puritans' view of their colony as a garden in the wilderness ordained the slaughter of native inhabitants; Frederick Law Olmsted expelled Irish and African American families to make way for a New York park to meet his liberal bourgeois ideal of democratic space.

The 1990 *Harper's* forum with which I introduced this book can be read critically in light of this problematic. Considering that Michael Pollan's interlocutors were all male, all white, and all American, the scope of the conversation was ambitiously universal: "Only Man's Presence Can Save Nature." Critics in gender studies and American studies alike had long before exposed the potential of this Adamic metaphor to justify destructive as well ameliorative transformations in the New World. Still, since the 1990s the notion that we should view the earth as a planetary garden and ourselves as its caretakers seems to be gathering wider support. At times it appears to reiterate the most traditional, patriarchal, Western, and Judeo-Christian images of humanity.

Paradoxically perhaps, advocates for the new planetary garden claim that this view of a fully inhabited earth is more just than the wilderness environmentalism in the United States and ecological imperialism abroad that preceded it. Already one large-scale U.S. garden manifesto, Emma Marris's 2011 *Rambunctious Garden: Saving Nature in a Post-Wild World*, has proposed a garden-centered paradigm for planetary environmental citizenship. Marris declares that "we must temper our romantic notion of untrammeled wilderness and find room next to it for the more nuanced notion of a global, half-wild rambunctious garden, tended by us" (2). Despite the overall polemical tone of her book, Marris's contribution is to show how conservation work and ecological research around the globe, from Hawaiian rainforests filled with invasive tree species to Poland's supposedly "primeval" Białowieża forest, already involve much historical compromise and cultivated nuance. The targets for Marris's book are ecosystem purists and dogmatic practitioners of ecological restoration, rather than the stereotypical corporate enemies of environmentalist writing. In a contemporary world characterized by radically uncertain and unstable climates, their search for historical baselines for restoring ecosystems, Marris argues, makes less sense than more dynamic and politically pragmatic approaches. In the place of mourning a vanished world or seeking to rebuild a problematic version of precolonial Eden, Marris would have us see the value of new and designer ecosystems as well as patches of hybrid wild at the edges of settlements.

Marris's "rambunctious" and planetary-scale garden resembles the ideas of the French landscape architect, novelist, and philosopher Gilles Clément, particularly his concept of the "planetary garden." His work is refreshingly informed by thinking through a typology of spaces,[3] a local, regional, and planetary awareness of natural processes, and a concrete attention to choices made by actual gardeners, including the choice not to act and let self-seeded plants grow out of place. In productive ways, Clément keeps in play elements of choice and chance in gardening within the contemporary context. He repudiates the "project of total control," which he associates with "radicals of ecology and those of nostalgia" (17). Like Marris, he sees little sense in practicing ecological restoration through wholesale herbicide extermination of invasive and nonnative plants. Restorationist "radicals" oppose the, to him, unavoidable quality of "nomadism" of plants through the dispersal of their seeds, particularly windborne seeds. Clément also imagines this nomadic quality as a principle not of entropy but of playful design. "Unthinkable landscapes are already being designed in the sky," he writes. "Were we to harvest the clouds, we would be surprised to find unpredictable seeds mixed with loess, fertile silt" (17). His attention here alternates between an atmospheric, planetary design and hyperlocal autonomy, which is an appropriately nomadic and planetary concept of gardening for the twenty-first century.

However, significant practical challenges, and perhaps unintended conceptual problems, are involved in materializing the planetary garden paradigm. The media coverage of the opening of the Svalbard Global Seed Vault in 2008, a scant four months before the onset of a global financial crisis, provides an apt example of the problematic nature of the planetary garden metaphor. Given that bulbs and seeds, as the first commodities, historically have fueled wild speculative markets (notably for tulip bulbs in Holland in the seventeenth century and for wheat on the Chicago Board of Trade in the nineteenth century), the coincidental mingling of metaphors in media coverage of the Svalbard seed bank now seems prophetic.

Elisabeth Rosenthal of the *New York Times* filed her February 29, 2008, report with the subtitle "A Fort Knox of Food." Three days earlier the newspaper's veteran climate reporter Andrew Revkin had asked more pointedly about this concentration of seed capital: "The groups funding the seed vault, including the Gates Foundation, say they are also pouring money into creating databases and other mechanisms for maintaining poor countries' access to the full array of crop strains. But what about the farmers in the field?" Through the past several years, while philanthropic foundations added digital entries to their global seed database, small-scale farmers who produced this vast reproductive wealth often protested the boosters of neoliberal economic globalization. Notwithstanding their protests and Revkin's pointed questions, thousands of small-scale cultivators from Africa, Mexico, Japan, and all points of the globe eventually sent more than three-quarter million varieties of food crop seeds to be frozen in the vault built beneath the permafrost of the Svalbard Archipelago.

This image of thousands of inert seeds frozen in perpetual winter for the good of posterity captures a millennial mood, but it also raises several questions.[4] How might projects of planetary cultivation help us cope with the uncertain threats but undeniably hard costs of global climate instability? How do we best direct our energies? Do we willingly participate, actively resist, or propose alternatives (such as the Bay Area Seed Interchange Library and similar local initiatives elsewhere) to a single world seed bank? The convergence of socioeconomic and ecological pressures in the twenty-first century promises that the Svalbard phenomenon will likely be repeated by other vaguely imperial yet seductively earthy planetary garden projects. It also underlines that, as helpful as it may be to critique narrow restoration goals, some work of ecological preservation may be necessary for environmental justice, and vice versa. In the Svalbard vault, preserving biodiversity in the form of seeds may overlap with future goals of maintaining fair access to food.

I conclude by considering two contemporary gardens rooted in the history this book has explored, and one ongoing national gardening project involving clones, to clarify how and why a convergence of ecological restoration and environmental justice is taking place. These final stories bring home the extent to which postindustrial gardens work responsively as a collective autobiographical form—or, in other words, how gardens made in a postindustrial society involve bumping up against the economic and ecological conditions for articulating self and community.

My first example comes from the Canadian philosopher and land manager Eric Higgs, who traces his own interest in ecological restoration to the eclectic home garden of Robert Dorney, one of his professors in Waterloo, Ontario, in the 1970s. Dorney, a professor of ecology and environmental planning, studied wildlife ecology with Aldo Leopold in Madison, Wisconsin, in the 1940s and settled in a suburban house near the University of Waterloo. Higgs relates how "to the chagrin of many neighbors," Dorney "ripped up his lawn and planted what he called a "miniecosystem," a small forest, prairie, and wetland in a hundredth of an acre." He may have been inspired by the experiments in restoring prairies and oak savannah in Madison. Higgs celebrates Dorney's willingness to experiment and his desire to maximize variety even within a small space, his piquant pleasure perhaps in nonconformity and sustaining life-long investment of energy. Dorney's untimely death—he died while cutting down an apple tree in 1987—led to the creation of a memorial garden in his name, which was also intended as a form of microrestoration (5). It reestablished a visual reference environment within the boundaries of campus while also making the public space more participatory—indeed, participation is mandatory or it will not remain a garden. The Dorney memorial campus garden, a territory of ideas and cultivation in process, involves staff who are serious gardeners and second-year philosophy majors with little prior garden experience or who may have grown up on a dairy farm, growing acres of vegetables in family gardens. Higgs describes how a well-cultivated garden can become a international and intergenerational dissemination of ideas and practices as well as of trees, plant stock, and seeds long after the gardener's death.

The philosophy of "nature by design" that Higgs advocates for ecological restoration is ultimately reconcilable with Marris's rambunctious garden and her preference for more openness and flexibility. Higgs recognizes an intrinsic value of wilder ecosystems at the ecological rather than individual level, and at the ecosystem rather than species or population level. This value system complements rather than devalues an emphasis within environmental justice on the radical equality and intrinsic value of human lives. Higgs calls ecological restoration "a way of conceptualizing different meanings for nature, a metaphor for appropriate intervention in natural processes, and an animating social idea" (7). Clearly, rather than a static antisocial purism, this version assigns a dual social and ecological value to the work—and its point of reference is gardening rather than wilderness.

The second example is the extensive landscape around Riverside Park in Milwaukee, two miles upstream from the working Port of Milwaukee, on Lake Michigan. The park flanks a mixed-income neighborhood located between the affluent and predominantly white northern suburbs, the campus of the public university to the east, and Milwaukee's commercial and financial district to the south. As the name suggests, the Milwaukee River borders the park to the west, separating it from the less affluent and majority African American neighborhoods that have been devastated by the loss of their economic base and, more recently, an epidemic of foreclosures.

In Riverside Park, and particularly in the several ongoing landscape designs of the Urban Ecology Center in its heart, urban gardening demonstrates the potential to join environmental justice and ecological restoration work, bridge racial and class divisions, and reach across generations and national borders. The Urban Ecology Center (UEC) was begun by a biology professor at the University of Wisconsin and fueled by local high school students and community leaders from both sides of the river. It gained national recognition in Richard Louv's book popularizing the value of contact with the natural world for childhood learning, *Last Child in the Woods,* and won local acclaim for its success in bringing back visitors to a neglected public space in the 1990s. The UEC has landscaped its grounds with deep-rooted prairie perennial grasses and flowers and established community gardens alongside the rails-to-trails Oak Leaf path that marks the eastern border of Riverside Park's wooded trails. These gardens flourish in the opening created by the railroad right-of-way turned into a paved path. Above it on the western bank, red oaks in a grove planned by Fredrick Law Olmsted for the original Riverside Park cast their afternoon shade over runners and cyclists on the path below. During flash flooding in the summer of 2010—which many blamed on global warming—the path below the UEC's perennial gardens remained above water while several inches of sewage-laced storm water washed over it to the north and south.

But to appreciate the enduring significance of Riverside Park as a gardened landscape, we have to walk a few hundred feet south and downhill to the river, where the UEC's staff forester oversees a major ecological restoration project, the Rotary Centennial Arboretum. Ecological purists might protest that a reference condition for this riparian area is too incomplete and that it represents novel design rather than a restoration. Emma Marris (and the UEC forester as well) might reply that the whole Riverside Park area represents a successful "rambunctious garden" managed for a variety of urban interests and amenities, including slowing runoff and encouraging more ecological variety than the previous monoculture of invasive reeds.

Should you borrow a kayak from the UEC and paddle downstream to where the Milwaukee River opens into Lake Michigan, you will find a literal island of restoration on Lakeshore State Park. As individual places or even aggregate acreage, perhaps these two projects are insignificant for the city's working poor. Lakeshore State Park and the Rotary Arboretum function within a changing metropolitan environment in which deindustrialization, redevelopment, and social movements have created more greenspaces but also persistent and racially segregated poverty. Still, dozens of community gardens seeded by Will Allen's Growing Power and others planted by members of Milwaukee Urban Gardens and the Victory Garden Initiative have knitted together routes for honeybees and butterflies, with the river's wooded banks serving as a corridor for coyote and deer.

In postindustrial American cities like Milwaukee, gardening for environmental justice and, increasingly, large-scale urban agriculture would redistribute power to more stakeholders in the community and help reshape the city's landscapes. The cumulative effect of existing gardens has been to blur divisions between human and nonhuman interests. This blurring is integral to the quality of place in many postindustrial urban gardens, which are habitat for food animals, pollinators, and native species planted to attract beneficial insects as well as models for new forms of land management and civic life. When I led groups of students along the river trails during the semester, and we read the data on *E. coli* in the river collected by citizen scientists and learned about persistent toxins in the soils, I came to see Riverside Park as a place that points toward but does not fully achieve environmental justice.

The socioeconomic divisions in the wider city of Milwaukee were simply too large to be resolved by even the most revolutionary urban farm or garden, although these places have become critical meeting points for imagining a transition to something better.

Gardeners also find themselves at the forefront of engaging climate change, as an ongoing national project shows. In 2012, the U.S. Department of Agriculture reissued its Plant Hardiness Zone Map (PHZM), based on average annual low temperatures from 1976 to 2005. Comparing its rainbow-colored isotherms with those on the 1990 USDA map seemed to contradict warming trends that North American gardeners have observed anecdotally over the past two decades. On its website, the USDA Agriculture Research Service cautioned readers that "changes in zones are not reliable evidence of whether there has been global warming," although they explained that the new map "is generally one half-zone warmer than the previous PHZM throughout much of the United States, as a result of a more recent averaging period (1974–1986 vs. 1976–2005)." In less than one generation, average temperatures and annual lows in the continental United States have warmed. Iconic southern crops such as peaches, bred for slightly more hardiness, are now more reliably harvested in the upper Midwest. Tender annuals such as tomatoes and peppers from year to year are set out long before traditional planting dates. (In Wisconsin, where I gardened until 2012, I started trying annuals outside closer to May Day than Memorial Day.)

Concurrent, then, with the growth of the national and international garden movements and expansion of narratives about environmental justice are murmurs of latent climate uneasiness. Such awareness among gardeners is discontinuous and incomplete but certainly merits further research. For cultural survival, refining phenology or logic of seasonal events—ecological timing—may become increasingly valuable as we face warmer and wilder seasons ahead. And more of us may need to learn from actual practices of cultivation, information sharing, and new garden media—perhaps more so than we need to adopt a new planetary garden paradigm or philosophy of ecological restoration.

In shifting our view of U.S. gardens to one of planetary climate futures instead of private or literary cultivation, it becomes sensible to emphasize new media tools, citizen science, and an emerging phenology of survival in our unstable, anthropogenic climate. The National Phenology Network has recruited thousands of home gardeners as citizen scientists gathering data on first leafing, bloom times, and insect appearances, which it collects in an online database and portal called Nature's Notebook. Hundreds of test gardens have been involved, many with data for over fifty years, led by two generations of ecologists at public universities in Nebraska, Montana, Wisconsin, Arizona, and elsewhere. Since the 1980s, moreover, the agricultural extension agencies of public universities have offered free, genetically identical common lilac plants to home gardeners willing to record annual leaf and bloom times. By planting cloned lilacs (and honeysuckle, until distribution of this invasive was discontinued), this national citizen-science project eliminates genetic variability in order to isolate the effects of climate. The results confirm widespread, dramatic, and everyday changes—the above-average warm spring of 2012 coincided with the earliest blooms for cloned lilacs on record, dating back to 1961.

The economic significance of understanding the impacts of shifting seasonal events on agriculture is one practical argument in favor of large-scale phenological study. It also opens important new avenues for historical research on gardens and garden writing, as scholars have sought useful data from the garden and field books of Jefferson, Thoreau, Leopold, and many others. By attending to the precarious and increasingly variable events of the plant world upon which other animals depend, contemporary gardeners also furnish a model of networked backyard-to-backyard cooperation that may become critical for a resilient future. This backyard and online phenology network requires regular and ongoing creative participation, a shuttling between social media and nature study. Rather than banking a maximum accumulation of only useful crop seeds and their wild progenitors—biodiversity under permafrost—the phenology network's data and gardens grow into the future.

Recuperating traditions of American garden writing from the previous century might also provide guideposts for sustaining a habitable planet. It is vital to further this work, for we live in an era of human-caused environmental degradation and economic divisions that cry out for alternative futures. One influential concept is the biologist Eugene Stoermer's *Anthropocene,* a term more often associated with his coauthor, the Nobel Prize–winning chemist Paul Crutzen and many others to describe the geologic-scale impact of human societies on planetary biophysical systems.[5] The Anthropocene concept imagines human history, its present and future, in terms parallel to the planet-as-garden metaphor; it carries its own ideological and normative risks (Malm and Hornborg 65). As Eileen Crist puts it: "If the Anthropocene's dream to avert scarcity for 10 billion humans (on a gardened smart planet) is somehow realized, scarcity will painfully manifest elsewhere—in homogenized landscapes, in emptied seas, in nonhuman starvations, in extinctions" (144).

We also live in a time of extreme economic inequality in which the top 1 percent of earners in the United States now garners nearly 20 percent of total earnings, which explains why other thinkers, including Jason Moore and Donna Haraway, suggest that "Capitalocene" may be a more fitting description of our era (159). Although the degree of inequality in the United States is exceptional when compared to Western Europe, Scandinavia, and Japan, data from the Organization for Economic Cooperation and Development indicate that economic inequality among OECD member nations has worsened globally in the past thirty years.[6] A national framing of environmental justice does not suffice in this planetary rift between widening inequalities and newly emerging discourses of universality.

Critical writers on gardening increasingly wrestle with this combination of economic, political, and ecological forces, but they can also draw on the resourceful ways in which writers over the past century have cultivated a broad and deep concept of environmental justice. They have done so by modeling weakly anthropocentric perspectives of the ecological conditions of gardens and a strong democratic and progressive awareness of the unequal politics of cultivation. Such garden writing has been at once literary, environmental, and political work. It has been the work of cultivating lives.

NOTES

Introduction

1. Anne Cary Randolph's household accounts are in *Thomas Jefferson Papers Series 7, Miscellaneous Bound Volumes,* available online in digital facsimile through the Library of Congress American Memory Project.

1. The Democratic Roots of Twentieth-Century U.S. Garden Writing

1. William Cronon describes this process of landscape transformation in New England in *Changes in the Land.* Alfred Crosby introduced the term *neo-Europes* in his classic global environmental history, *Ecological Imperialism,* arguing that disease and domestic animals acted as coinvading forces in the Americas and other temperate latitudes where European colonizers settled. Crosby's biogeographical thesis has been both built upon and contested by more recent work focusing on regions outside North America; see, for example, Cushman, *Guano and the Opening of the Pacific World;* and Mavhunga, *Transient Workspaces.*

2. Jefferson's correspondence with M'Mahon includes over 40 letters between 1806 and 1815, most of which were written between 1806 and 1809, concerning M'Mahon's role as custodian of the botanical specimens collected by the Lewis and Clark expedition.

3. Environmental historians, beginning with Carruthers's 1995 history of the Kruger National Park in South Africa and then William Cronon's "Trouble with Wilderness" (1996) and Karl Jacoby's *Crimes against Nature* (2005), have systematically exposed the seemingly natural wilderness as political, naturalized, but designed landscapes as well as territories of real human dispossession with deliberately invented traditions of wildness.

4. Of these three, the least-known but still noteworthy is German prince Hermann von Pückler-Muskau's estate, which he designed and used as the basis for his theory of landscape gardening published in 1834. Pückler's work was translated and

circulated widely in England and America, where his influence is most visible in the design of Olmsted's Biltmore Estate in North Carolina.

5. The relationship between Late Victorian garden writing and U.S. cultural elites is exemplified by *Century* magazine commissioning the novelist Edith Wharton to write a series of essays, published as *Italian Villas and Their Gardens* in 1904.

6. On the political and social context of the development of ecological science by Cowles and colleagues in geography at the University of Chicago, see Mitman, *The State of Nature*.

7. In chapter 3, I explore the idea that gardening could cross social boundaries in the context of racial and regional divisions in southern gardens.

8. See Jordan and Lubick, *Making Nature Whole*, 30. Two recent histories of the development of ecological restoration date the practice to the 1930s. Jordan and Lubick identify four projects in the United States and two in Australia, where restoration practices seem to have emerged concurrently and independently through local scientists, gardeners, and conservationists attempting to counteract soil degradation, deforestation, and species loss in local landscapes.

9. See, for example, Lekan, "*Serengeti Shall Not Die.*"

10. In *American Georgics* (2001), Sweet is interested in a general structure of environmental alienation represented in Americans' experience of nature "primarily in the pastoral mode, regarding nature (if it is regarded at all) as a site of leisure, not of labor" (175). The term *environmental alienation* echoes two moments in Marx's concept of alienation: on the one hand, the necessary process whereby people convert "first nature" to usable "second nature" in the form of both commodities and ideas; on the other hand, the reified social relations of capitalism that separate wage workers from nature as a means of production and as part of human species life.

11. Wendell Berry's *Unsettling of America* (1977) reads as a prophetic study of the consequences of agricultural concentration. For a critical introduction to studies of the U.S. food system, see Heffernan, "Concentration of Ownership and Control in Agriculture"; Julie Guthman's work on the political economy of organic agriculture in *Agrarian Dreams;* and Bell, *Farming for Us All*.

12. The environmental historian Donald Worster has provided one of the most compelling accounts of the dust bowl as the result of Great Plains ecology, capitalist economy, and an all-too-human but thoroughly American culture of heedless optimism in *Dust Bowl*. The more recent project "Rethinking the Dust Bowl, 1830–1941," led by Geoff Cunfer, incorporates digital and GIS methods and offers a deeper chronology (www.hgis.usask.ca).

13. See esp. Phillips, *This Land, This Nation;* and Conlogue, *Working the Garden*.

14. Such manuals appear and are often republished during recurrent economic crises; for example, *Ten Acres Enough* by Edmund Morris (1864; republished in 1905, 1996, and 2004); *Three Acres and Liberty* by Bolton Hall (1907, 1917, and 1922); and M. G. Kains's *Five Acres and Independence* (published in 1935, with a new edition in 1945; reprinted in 1973).

15. A genealogy of cultural critique in this vein includes Paul and Percival Goodman's

Communitas (1948), Lewis Mumford's *Culture of Cities* (1938), and E. F. Schumacher's *Small Is Beautiful* (1973).

16. Expanded suburban development is the logical extension of Borsodi's homestead ideology, the unintended consequence of garden ideology meeting the federally insured 30-year mortgages of returning veterans. For a history of suburban development in the U.S., see Jackson, *Crabgrass Frontier*.

17. These included small farmers and large farmers and ranchers; a range of farm organizations and product corporations; the fragile Farm Security Administration and a burgeoning USDA; the War Food Administration and Commodity Credit Program, which managed agricultural price supports; and a legion of writers, editors, home gardeners, and farmers who debated the consequences of the politics of agriculture. The fading significance of subsistence or marginal farming in the politics of U.S. agriculture has been attributed to a shift in the balance of power between these various groups, specifically to the ascendancy of agribusiness interests. The classic polemic on this shift is Wendell Berry's 1977 *The Unsettling of America*, although recent historians have examined in greater detail how New Deal programs set postwar agricultural policy with more complex unintended consequences.

18. The singular exception is J. I. Rodale's *Organic Farming and Gardening* (later simply *Organic Gardening*), published by his independent press in Emmaus, Pennsylvania. See Andrew Case's history (forthcoming) of Rodale and organic gardening in the United States.

19. Cecilia Gowdy-Wygant argues that particularly in the U.S., victory gardens and the Women's Land Army of America created opportunities for broader civic roles for women while also framing a context for redomesticating women in the postwar era—subsuming gardening within the ambit of middle-class kitchens and consumption (5).

20. On literary responses to rationing and war austerity in food writing at midcentury, see Carruth, *Global Appetites*, ch. 3.

2. Postwar Garden Writing, Literary Cultivation, and Environmentalism

1. The U.S. Food and Drug Administration reports on pesticide residues found in food, which represents only a fraction of the total annual biological burden of trace amounts of lethal chemicals that U.S. residents absorb through their air, water, and soil. Nonetheless, the agency's annual reports since 1987, filed online under its Pesticide Program Residue Monitoring, makes for startling reading and a contemporary counterpoint to Carson's text.

2. See Robbins and Birkenholtz, "Turfgrass Revolution"; Robbins's *Lawn People* builds on a decade of studies of American lawn culture and national homeowner surveys and argues that U.S. chemical lawn applications are the result of politics and economics rather than "lawn culture." The author argues that lawn care and chemical manufacturers, particularly since the 1980s, have successfully marketed lawn chemical applications to Americans to meet production increases that compensate for falling per-unit prices (93).

3. The most extensive data on annual lawn applications of pesticides and top-selling chemicals, and their toxicity, is available from the California Pesticide Information Portal (CALPIP). The FDA Pesticide Program Residue Monitoring published fiscal year reports on the presence of trace pesticides in foods purchased from U.S. supermarkets, but often with a multiyear delay between data collection and publication online (e.g., the 2012 report was released Feb. 2015). In the 2012 Pesticide Report, 70.5 percent of domestic fruit samples were found to contain pesticide residues, with another 1.5 percent containing residues in violation of EPA tolerance (118).

4. See Jackson, *Crabgrass Frontier*; Phillips, *This Land, This Nation*, esp. ch. 4.

5. See also Gowdy-Wygant, *Cultivating Victory*, for a historical analysis of the persistent influence of nostalgic, nationalistic conceptions of the victory garden and the almost complete forgetting of the agricultural and cultural work of the Women's Land Armies.

6. For an insightful study of the Nearings' work and the broader context of homesteading traditions as "lived religion" in America, see Gould's *At Home in Nature* (6). Gould worked closely with Helen Nearing toward the end of Helen's life and conducted research for her book at the Nearings' Good Life Center in Maine. Gould's argument about the Nearings' homesteading in particular—that Scott Nearing was as motivated by spiritual as political and material analysis—jibes more with Helen Nearing's interpretation of Scott in *Loving and Leaving the Good Life* than his self-portrayal in his autobiography as vehemently anticlerical and anti-established religion.

7. Scott Nearing lectured widely in the heyday of American socialism, when he befriended Upton Sinclair, ran for Congress on the Socialist Party ticket against Fiorello La Guardia, and corresponded with Eugene Debs; see *The Making of a Radical*, 176.

8. Samuel Hays has argued that dominant attitudes toward the environment shifted in the early postwar decades from a conservation focus on nature as a resource to be managed to an environmentalist approach focused on nature as consumer amenity and healthful environment, the values highlighted by the title of his study: *Beauty, Health, and Permanence*.

9. The *St. Nicholas* children's magazine was an important vehicle that inspired a remarkable number of professional writers of Rachel Carson's and E. B. White's generation; it also regularly published essays on nature study and appreciation. On the particularly strong influence of the nature study movement on children of the Progressive Era, including Rachel Carson and Aldo Leopold, see Armitage, *The Nature Study Movement*, 210.

3. *Being There*, *Second Nature*, and the Gardener as Pragmatist

1. The most persuasive and historically grounded study of the linkage of property and white privilege, which results in a discounted set of psychological benefits in the U.S. over the twentieth century—is George Lipsitz's *Possessive Investment in Whiteness* (2000).

2. On the importance of emotions in responding to environmental issues, see especially Kari Norgaard, *Living in Denial: Climate Change, Emotions, and Everyday Life* (2011). The biologist and writer Elin Kelsey has argued increasingly that scholars and activists must move beyond the "doom and gloom" narrative focusing on environmental degradation. See the contributions in the volume Kelsey edited, "Beyond Doom and Gloom: An Exploration through Letters," *RCC Perspectives* no. 6 (2014).

3. The object of Pollan's anecdote of the "gift"—finding a large Sibley squash hidden in his garden late in the season—feels more specific and significant. The Sibley is not a fetish or pure aesthetic object nor a mass-market pumpkin, but rather a filling banana-type squash famed for its flavor and with a rich history of cultural exchange, some of which Pollan hints at in *Second Nature*. Introduced in the late eighteenth century to Massachusetts from the Andes, it quickly spread to gardeners among the Arikara and Winnebago nations along the Missouri River, then was "rediscovered" by Euro-American settlers in Iowa and marketed in the mid-nineteenth century by a commercial seedsman, Hiram Sibley of New York. See Gary Nabhan, *Renewing America's Food Traditions*, 60.

4. The critic Annette Kolodny has described such imagery as central to foundational myths of American settlement in her series of books, most dramatically in *The Lay of the Land*.

4. Race, Regionalism, and the Emergence of Environmental Justice in Southern Gardens

1. On the systematic effects of institutional racism on black farmers in the South, see Green, Green, and Kleiner, *Cultivating Food Justice*, esp. 57–60; on the many resonances of "black hunger" in U.S. culture, see Witt, *Black Hunger*. Warnes's *Hunger Overcome?* offers a particularly strong reading of literary representations of food consumption and production in the South that in many ways runs parallel to my arguments focusing on gardens and cultivation while drawing from a different archive.

2. See Stewart, "What Nature Suffers to Groe," and "Rice, Water, and Power," his study of enslaved peoples' strategies of resistance, including gardening in the margins of the hydraulic plantation landscape of the Lowcountry. In a garden tour of Rosedown Plantation Historic Site in St. Francisville, La., the historical curator Trish Aleshire emphasized that the white mistress relied on skilled gardeners to maintain her expansive botanical collections. Former plantations like Rosedown were converted to botanical gardens across the South in the early twentieth century. Several have since been opened to the public, including Brookgreen Gardens in Murrells Inlet, South Carolina, and Maclay Gardens (State Park) in Tallahassee, Florida.

3. A number of flowers are stereotypically "southern," such as jessamine, honeysuckle, and especially magnolia. Thus the whiteness of flowers hardy only in southern states inadvertently signifies race and gender relations. The garden writer Eleanor Perényi also pointed out the correlation of pure white flowers and class pretensions in her book *Green Thoughts* (1981). Black writers have signified on the ironies of the luxuriant production of whiteness (cotton bolls, magnolia blooms) from the blackness

of southern soils, for example W. E. B. Du Bois's *Quest for the Silver Fleece* (1911) and Zora Neale Hurston's *Their Eyes Were Watching God* (1937).

4. Joseph Bodziock has argued that Douglass also relies on gothic conventions to undermine an eighteenth- and nineteenth-century trope of the South as a realized Garden of Eden in an effort to connect with and persuade his Northern audiences (253).

5. In *My Bondage and My Freedom* (1855), Douglass emphasizes the pleasure he took in the woods and wildlife surrounding Lloyd's plantation. He implies that the enjoyment of nature is a universal faculty and one that initially resisted the alienating effect of enslavement: "The tops of the stately poplars were often covered with the red-winged black-birds, making all nature vocal with the joyous life and beauty of their wild, warbling notes. These all belonged to me, as well as to Col. Edward Lloyd, and for a time I greatly enjoyed them" (68).

6. This value of collecting plant material for materializing community memory is both shared across region, class, and racial groups and particular to individual gardens as recorded in southern fiction. See Harrison's *Female Pastoral* (esp. 12); Patricia Yaeger, *Dirt and Desire;* and Gundaker's studies of vernacular gardens as home grounds in *Keep Your Head to the Sky* and, with Judith McWillie, the fascinating set of interviews and photographs in *No Space Hidden.*

7. Glave, in "'A Garden So Brilliant with Colors,'" describes interracial garden competitions sponsored by agricultural agencies during the Progressive Era, which presents a benign contrast to the discriminatory practices of agricultural lending through county commissioners at the end of the twentieth century; these were exposed by the *Pigford v. Glickman* class-action lawsuit over racial discrimination in federal loans to farmers.

8. The relevant passage is also one of the most often cited from *Black Elk Speaks;* it follows Black Elk's description of the Wounded Knee Massacre of 1890: "I did not know then how much was ended. When I look back now from this high hill of old age, I can still see the butchered women and children lying heaped and scattered along the crooked gulch as plain as when I saw them with eyes still young. And I can see that something else died there in the bloody mud, and was buried in the blizzard. A people's dream died there. It was a beautiful dream . . . the nation's hoop is broken and scattered. There is no center any longer, and the sacred tree is dead" (218).

9. See Robert Bullard and Beverly Wright, eds., *Race, Place, and Environmental Justice after Hurricane Katrina;* and Stephanie Lemenager, "Petro-Melancholia." George Beckford, in *Persistent Poverty,* has classified the American South with countries having plantation legacies and underdevelopment.

5. Postindustrial America and the Rise of Community Gardens

1. The opposition movement against the Lancer incinerator by Latina and African American women in South Central became one of the paradigmatic cases for theorizing environmental justice, beginning with Robert Bullard's discussion of the

case in *Confronting Environmental Racism* (1993), 35. See also Laura Pulido's essay "Community, Place, and Identity" (1997), based on fieldwork with activists in South Central, in which she argues for a dynamic understanding of place-based identity and environmental justice activism suited to multiethnic social networks in U.S. culture.

6. Seeding New Territories

1. On these two dominant narrative structures in U.S. environmental history, see William Cronon, "A Place for Stories: Nature, History, and Narrative."

2. The historian Patricia Seed draws a similar conclusion about the role of English botanical collecting and gardening in the Caribbean in a comparative context with Spanish and French colonialism in the Americas; see Seed, *Ceremonies of Possession*.

3. By now the body of scholarship dedicated to critiques of the hegemonic power of nature in American culture is too large to summarize. Early salvos include feminist work by Carolyn Merchant and Greta Gaard (1994) through more recent books such as Paul Outka's *Race and Nature from Transcendentalism to the Harlem Renaissance* (2008), Dianne Glave's *Rooted in the Earth: Reclaiming the African American Environmental Heritage* (2010), Kimberly Ruffin's *Black on Earth: African American Ecoliterary Traditions* (2010), and Nicole Seymour's *Strange Natures: Futurity, Empathy, and the Queer Environmental Imagination* (2013). A common strategy of these authors is to identify alternative histories and traditions of working with and forming connections to the natural environment among subjugated communities, arguing for the ecological as well as social value in their wider recognition.

4. The urban agriculture movement is by no means limited to North America and Western Europe. See for example Andrea Gaynor, *Harvest of the Suburbs: An Environmental History of Growing Food in Australian Cities* (2006). A 2012 project, "Urban Farming and the Agricultural Show," sponsored by the National Museum of Australia and the University of Canberra, focused on the growing movement in urban farming and suburban backyard gardening and connected it through interviews with gardeners who exhibit their produce at annual agricultural shows across Australia. A Canadian small business called "The Urban Farmer" leads tours and organizes "sustainable urban agriculture and permaculture internships" on urban farms in Havana, Cuba, a country that, since the fall of the Soviet Union, has by necessity become a world leader in urban agriculture.

5. Steinberg's term *organic city* may confuse those familiar with Emile Durkheim's notion of *organic solidarity*, a term often used to analyze overlapping historical phenomena of class differences, work, and industrialization. Durkheim argued in his influential book *The Division of Labor* (1893) that urban and rural dwellers in industrial societies had become increasingly interdependent through a division of specialized labor, replacing the "mechanical" connections of traditional societies with "organic" connections.

6. The USDA's working group simplified this definition in 2010 for its Food Access Research Atlas (an online mapping tool formerly called the Food Desert Locator) as

a low-income census tracts with "low access to a supermarket or large grocery store" (www.ers.usda.gov/data-products).

Epilogue: Garden Writing and a Phenology of Survival

1. For a contextualization of Michelle Obama's garden in terms of gender roles and the national cult of victory gardening, see Gowdy-Wygant, *Cultivating Victory*, 182–84; in terms of the race and class dimensions of the discourse of food justice in tension with the ideology of obesity, see Guthman, *Weighing In*, 185–87.

2. The American Community Garden Association's website has a wealth of information on the state of the community garden movement nationally, including a database of gardens across the country and survey data on the benefits such gardens bring to neighborhoods. See also Flores's *Food Not Lawns*. I cannot do justice here to the international organic gardening and slow food movements, which have been deeply critical of status quo agriculture and environmental politics; see esp. Rodale Press publications, the *Slow Food* newsletter, and the *Snail*, as well as Dan Philippon's ongoing research, an overview of which he has presented online via the Rachel Carson Center.

3. Clément seems to have borrowed his typology and perhaps some of his inspiration from the work of the German urban ecologists Herbert Sukopp and Ingo Kowarik, who developed their ideas and practices using Berlin as an enclosed laboratory during the Cold War; see Lachmund, 47–81.

4. The literary critic Reinhard Hennig is currently at work on a project analyzing the many rhetorical framings of the opening of the Svalbard Global Seed Vault, including utopian, apocalyptic, and Norwegian nationalist narratives about the secured site.

5. For critical engagements with the concept of the Anthropocene, see Haraway; Malm and Hornborg; Crist; and Trischler et al.

6. The 2011 report *Divided We Stand* finds that average incomes of the richest 10 percent of Americans were roughly 14 times those of the poorest 10 percent, compared to an OECD average ratio of 9 to 1. Among developed Western nations, the United States led this trend toward growing extreme inequality.

WORKS CITED

Adamson, Joni. *American Indian Literature, Environmental Justice, and Ecocriticism: The Middle Place.* Tucson: U of Arizona P, 2001. Print.

Adamson, Joni, Mei Mei Evans, and Rachel Stein, eds. *The Environmental Justice Reader: Politics, Poetics, & Pedagogy.* Tucson: U of Arizona P, 2002. Print.

Aleshire, Trish. Garden tour. Rosedown Plantation Historic Site. St. Francisville, LA. 2 Mar. 2007.

Alkon, Alison Hope, and Julian Agyeman, eds. *Cultivating Food Justice.* Cambridge: MIT P, 2011. Print.

Allen, Barbara L. *Uneasy Alchemy: Citizens and Experts in Louisiana's Chemical Corridor Disputes.* Cambridge: MIT P, 2003. Print.

Allen, Will, and Charles Wilson. *The Good Food Revolution: Growing Healthy Food, People, and Communities.* New York: Gotham, 2013. Print.

Anderson, Amanda. *The Way We Argue Now: A Study in the Cultures of Theory.* Princeton: Princeton UP, 2006. Print.

Angell, Katharine S. "Home and Office." *Survey Graphic* 57 (1926): 318–20. Print.

"Anthropocene: Envisioning the Future of the Age of Humans." Ed. Helmuth Trischler. *RCC Perspectives* 3 (2013). Print.

Armitage, Kevin. *The Nature Study Movement: The Forgotten Popularizer of America's Conservation Ethic.* Lawrence: U of Kansas P, 2009. Print.

Azima, Rachel Mieko Pouri. *Alien Soil: Ecologies of Transplantation in Contemporary Literature.* Diss. U of Wisconsin, 2008. Ann Arbor: UMI, 2008. ATT 3327781. Print.

Bailey, Liberty Hyde. *The Holy Earth.* New York: Scribner's, 1915. Print.

———. *The Survival of the Unlike.* 6th ed. New York: Macmillan, 1911. Print.

Balmori, Diana, and Margaret Morton. *Transitory Gardens, Uprooted Lives.* New Haven: Yale UP, 1993. Print.

Beckford, George. *Persistent Poverty: Underdevelopment in Plantation Economies of the Third World.* 2nd ed. Kingston, Jamaica: U of the West Indies P, 1999. Print.

Beecher, Catharine E., and Harriet Beecher Stowe. *The American Woman's Home.* New York: Ford, 1869. Print.

Being There. Dir. Hal Ashby. Perf. Peter Sellers. Screenplay by Jerzy Kosinski and Robert C. Jones. United Artists, 1979. Film.

Belasco, Warren J. *Appetite for Change: How the Counterculture Took on the Food Industry*. Ithaca, NY: Cornell UP, 1993. Print.

Bell, Michael. *Farming for Us All: Practical Agriculture and the Cultivation of Sustainability*. University Park: Pennsylvania State UP, 2004. Print.

Bennett, Michael. "Anti-Pastoralism, Frederick Douglass, and the Nature of Slavery." *Beyond Nature Writing*. Ed. Karla Armbruster and Kathleen R. Wallace. Charlottesville: U of Virginia P, 2001. 195–210. Print.

Bergthaller, Hannes, et al. "Common Ground: Ecocriticism, Environmental History, and the Environmental Humanities." *Environmental Humanities* 5 (2014): 261–76. Web. 13 Nov. 2014.

Berkes, Fikret. *Sacred Ecology*. 2nd ed. New York: Routledge, 2008. Print.

Berry, Wendell. *The Hidden Wound*. Boston: Houghton, 1970. Print.

———. *The Unsettling of America: Culture and Agriculture*. San Francisco: Sierra Club, 1977. Print.

Beverley, Robert. *The History and Present State of Virginia, in Four Parts*. London: R. Parker, 1705. *Documenting the American South*, 2006. University Library, UNC. 24 June 2009. Web. 21 Mar. 2012.

Black Elk, John G. Neihardt, and Raymond J. DeMallie. *Black Elk Speaks: Being the Life Story of a Holy Man of the Oglala Sioux*. Albany: State Univ. of New York P, 2008. Print.

Bodziock, Joseph. "In the Cage of Obscene Birds: The Myth of the Southern Garden in Frederick Douglass's *My Bondage and My Freedom*." *The Gothic Other: Racial and Social Constructions in the Literary Imagination*. Ed. Ruth Bienstock Anolik and Douglas L. Howard. Jefferson, NC: McFarland, 2004. 251–64. Print.

Booth, Wayne. *The Company We Keep: An Ethics of Fiction*. Berkeley: U of California P, 1988. Print.

Borsodi, Ralph. *Flight from the City*. New York: Harper, 1933. Print.

Bourdieu, Pierre. *Distinction: A Social Critique of the Judgment of Taste*. Trans. Richard Nice. Cambridge: Harvard UP, 1984. Print.

———. *Language and Symbolic Power*. Cambridge: Harvard UP, 1999. Print.

Brandt, Deborah. *Literacy in American Lives*. New York: Cambridge UP, 2001. Print.

Buell, Laurence. *The Future of Environmental Criticism: Environmental Crisis and Literary Imagination*. Malden, MA: Blackwell, 2005. Print.

———. *Writing for an Endangered World: Literature, Culture, and Environment in the U.S. and Beyond*. Cambridge: Harvard UP, 2001. Print.

Bullard, Robert D. *Confronting Environmental Racism: Voices from the Grassroots*. Boston: South End, 1993. Print.

Bullard, Robert D., and Beverly Wright, eds. *Race, Place, and Environmental Justice after Hurricane Katrina: Struggles to Reclaim, Rebuild, and Revitalize New Orleans and the Gulf Coast*. Boulder, CO: Westview, 2009. Print.

Cable, George Washington. *The Amateur Garden*. New York: Scribner's, 1914. Print.

Čapek, Karl. *The Gardener's Year*. 1931. Trans. M. and R. Weatherall. Madison: U of Wisconsin P, 1984. Print.
Carney, Judith A., and Richard Nicholas Rosomoff. *In the Shadow of Slavery: Africa's Botanical Legacy in the Atlantic World*. Berkeley: U of California P, 2010. Print.
Carpenter, Novella. *Farm City: The Education of an Urban Farmer*. New York: Penguin, 2009. Print.
Carruth, Allison. *Global Appetites: American Power and the Literature of Food*. Cambridge: Cambridge UP, 2013. Print.
Carruthers, Jane. *The Kruger National Park: A Social and Political History*. 1995. Scottsville, South Africa: U of KwaZulu-Natal P, 2013. Print.
Carson, Rachel. *Silent Spring*. 1962. Boston: Houghton, 1990. Print.
Chivers, C. J. "After Uprooting Gardeners, City Razes a Garden." *New York Times*. 16 Feb. 2000. Web. 2 July 2007.
Christy, Liz. Interview. *Liz Christy Community Garden*. N.d. Web. 21 Oct. 2007.
Clayton, Virginia Tuttle. "Wilde Gardening and the Popular American Magazine, 1890–1918." *Nature and Ideology: Natural Garden Design in the Twentieth Century*. Ed. Joachim Wolschke-Bulmahn. Washington, DC: Dumbarton Oaks, 1997. 131–54. Print.
Clément, Gilles. *Planetary Gardens: The Landscape Architecture of Gilles Clément*. Ed. Alessandro Rocca. Boston: Birkhäuser, 2008. Print.
Cobbett, William. *The American Gardener: A Treatise on the Situation, Soil, and Laying Out of Gardens, on the Making and Managing of Hot-Beds and Green-Houses, and on the Propagation and Cultivation of the Several Sorts of Vegetables, Herbs, Fruits, and Flowers*. J. L. Gihon, 1854. *Making of America Project*. U of Michigan P. Web.
Cohen, Lizabeth. *A Consumers' Republic: The Politics of Mass Consumption in Postwar America*. New York: Vintage, 2004. Print.
Cohen, Michael P. "Blues in the Green: Ecocriticism under Critique." *Environmental History* 9.1 (2004): 9–36. Print.
Cole, Karen. "Tending the Southern Vernacular Garden: Elizabeth Lawrence and the Market Bulletin." *Such News of the Land: U.S. Women Nature Writers*. Ed. Thomas S. Edwards and Elizabeth A. De Wolfe. Hanover, NH: UP of New England, 2001. 160–69. Print.
"Community Garden Search." *Open Space Greening Program*. Council on the Environment of New York City, n.d. Web. 10 Oct. 2007.
Conlogue, William. *Working the Garden: American Writers and the Industrialization of Agriculture*. Chapel Hill: U of North Carolina P, 2001. Print.
Cooper, David E. *A Philosophy of Gardens*. Oxford: Oxford UP, 2006. Print.
Court, Franklin E. *Pioneers of Ecological Restoration: The People and Legacy of the University of Wisconsin Arboretum*. Madison: U of Wisconsin P, 2012. Print.
Cresswell, Tim. *Place: An Introduction*. 2nd ed. Malden, MA: Blackwell, 2015. Print.
Crevecoeur, J. Hector St. John. *Letters from an American Farmer and Sketches of Eighteenth-Century America*. Ed. Albert E. Stone. New York: Penguin, 1981. Print.

Crist, Eileen. "On the Poverty of Our Nomenclature." *Environmental Humanities* 3 (2013): 129–47. Web.
Cronon, William. *Changes in the Land: Indians, Colonists, and the Ecology of New England*. New York: Hill and Wang, 1983. Print.
———. "A Place for Stories: Nature, History, and Narrative." *Journal of American History* 78.4 (Mar. 1992): 1347–76. Print.
———. "The Trouble with Wilderness." *Uncommon Ground: Rethinking the Human Place in Nature*. New York: Norton, 1996. Print.
Crosby, Alfred W. *Ecological Imperialism: The Biological Expansion of Europe, 900–1900*. Cambridge: Cambridge UP, 1986. Print.
"Crowd Storms Former Garden to Protest Bulldozing by the City." *New York Times*. 6 Mar. 2000. Web. 2 July 2007.
Curtis, John. *Vegetation of Wisconsin: An Ordination of Plant Communities*. Madison: U of Wisconsin P, 1959. Print.
Cushman, Gregory T. *Guano and the Opening of the Pacific World: A Global Ecological History*. Cambridge: Cambridge UP, 2013. Print.
Davis, Linda H. *Onward and Upward: A Biography of Katharine S. White*. New York: Harper, 1987. Print.
Dawson, Laura. *City Bountiful: A Century of Community Gardening in America*. Berkeley: U of California P, 2005. Print.
"Death of a Garden." Narr. Jose Torres. *All Things Considered*. Natl. Public Radio. 24 Feb. 2000. Radio.
Deloughrey, Elizabeth. "Island Ecologies and Caribbean Literatures." *Tijdschrift voor Economische en Sociale Geografie* 95.3 (2004): 298–310. Print.
Denning, Michael. *The Cultural Front: The Laboring of American Culture in the Twentieth Century*. New York: Verso, 1996. Print.
Didur, Jill. "Strange Joy." *Interventions: International Journal of Postcolonial Studies* 13.2 (2011): 236–55. Print.
Dillon, Julia Lester. *The Blossom Circle of the Year in Southern Gardens*. New York: De La Mare, 1922. Print.
Dimock, Wai-Chee. *Residues of Justice: Literature, Law, Philosophy*. Berkeley: U of California P, 1996. Print.
Douglass, Frederick. *My Bondage and My Freedom*. 1855. New York: Miller. *Electronic Text Center*. n.d. Web. 12 May 2009.
———. *Narrative of the Life of Frederick Douglass, an American Slave*. 1845. *The Classic Slave Narratives*. Ed. Henry Louis Gates Jr. New York: Penguin, 2002. 323–436. Print.
Dowie, Mark. *Conservation Refugees: The Hundred-Year Conflict between Conservation and Native Peoples*. Cambridge: MIT P, 2009. Print.
Downing, Andrew Jackson. *Rural Essays*. New York: Leavitt, 1856. Print.
Eine andere Welt ist pflanzbar! / Another World Is Plantable! Community Gardens in North America Part 4. Dir. Ella von der Haide. 2012. Film.
Empson, William. *Some Versions of Pastoral*. London: Chatto, 1950. Print.

"Environmental Justice." US Environmental Protection Agency. Web. 1 May 2008.
Feenstra, Gail. "Creating Space for Sustainable Food Systems: Lessons from the Field." Agriculture and Human Values 19 (2002): 99–106. Print.
Fitch, Robert. The Assassination of New York. New York: Verso, 1993. Print.
Flores, Heather C. Food Not Lawns: How to Turn Your Yard into a Garden and Your Neighborhood into a Community. White River Junction, VT: Chelsea Green, 2006.
Franzen, Jonathan. The Corrections. New York: Farrar, 2001. Print.
———. "Why Bother?" How to Be Alone. New York: Farrar, 2002. 55–97. Print.
Freyfogle, Eric T. "Conservation and the Lure of the Garden." Conservation Biology 18.4 (2004): 995–1003. Print.
The Garden. Dir. Scott Hamilton Kennedy. Oscilloscope Pictures, 2009. Film.
Garmey, Jane, ed. The Writer in the Garden. Chapel Hill: Algonquin, 1999. Print.
Gaynor, Andrea. Harvest of the Suburbs: An Environmental History of Growing Food in Australian Cities. Crawley: U of Western Australia P, 2006. Print.
Georgia Farmer's Market Bulletin 40.3 (Aug. 1955). Atlanta: Bureau of Markets. Print.
Glave, Dianne D. "A Garden So Brilliant with Colors, So Original in Design: Rural African American Women, Gardening, Progressive Reform, and the Foundation of an African American Environmental Perspective." Environmental History 8.3 (2003): 395–411. Print.
———. Rooted in the Earth: Reclaiming the African American Environmental Heritage. 1st ed. Chicago: Lawrence Hill, 2010. Print.
Gold, Michael. Jews without Money. 1930. New York: Avon, 1965. Print.
Goodman, Kenneth Sawyer. 1914. "A Masque" [called variously "The Beauty of the Wild" or "At the Edge of the Woods"]. Morton Arboretum Quarterly 7.1 (1971): 9–15. Print.
Gottlieb, Robert, and Andrew Fischer. "'First Feed the Face': Environmental Justice and Community Food Security." Antipode 28.2 (1996): 193–203. Print.
Gould, Rebecca. At Home in Nature: Modern Homesteading and Spiritual Practice in America. Berkeley: U of California P, 2005. Print.
Gowdy-Wygant, Cecilia. Cultivating Victory: The Women's Land Army and the Victory Garden Movement. Pittsburgh: U of Pittsburgh P, 2013. Print.
Grampp, Christopher. From Yard to Garden: The Domestication of American Home Grounds. Chicago: Center for American Places, 2008. Print.
Grant, Greg. "Martha Turnbull." The Influence of Women on the Southern Landscape. Proc. of Tenth Conference on Restoring Southern Gardens and Landscapes, Oct. 5–7. Winston-Salem: Old Salem, 1995. Print.
Green, John J., Eleanor M. Green, and Anna M. Kleiner. "From the Past to the Present: Agricultural Development and Black Farmers in the American South." Alkon and Agyeman 47–64. Print.
Grese, Robert E. Jens Jensen: Maker of Natural Parks and Gardens. Baltimore: Johns Hopkins UP, 1992. Print.
Guha, Ramachandra. "Radical American Environmentalism and Wilderness Preservation: A Third World Critique." Environmental Ethics 11.1 (1989): 71–83. Print.

Gundaker, Grey. "Introduction: Home Ground." *Keep Your Head to the Sky: Interpreting African American Home Ground.* Ed. Grey Gundaker. Charlottesville: U of Virginia P, 1998. 3–23. Print.

Gundaker, Grey, and Judith McWillie. *No Space Hidden: The Spirit of African American Yard Work.* Knoxville: U of Tennessee P, 2005. Print.

Guthman, Julie. *Agrarian Dreams: The Paradox of Organic Farming in California.* 2nd ed. Berkeley: U of California P, 2014. Print.

———. *Weighing In: Obesity, Food Justice, and the Limits of Capitalism.* Berkeley: U of California P, 2011. Print.

Guthrie, Elizabeth, and Sara Levine. "Interview with a Victory Gardener." 1989. *Community Greening Review: 25 Years of Community Gardening.* New York: Amer. Community Garden Assoc., 2005. N.pag. Print.

Hall, Bolton. *Three Acres and Liberty.* 1907. New York: Macmillan, 1922. Print.

Hanchett, Thomas W. *Sorting Out the New South City: Race, Class, and Urban Development in Charlotte, 1875–1975.* Chapel Hill: U of North Carolina P, 1998. Print.

Haraway, Donna. "Anthropocene, Capitalocene, Plantationocene, Chthulucene: Making Kin." *Environmental Humanities* 6 (2015): 159–65. Web. 4 June 2015.

Harrison, Elizabeth Jane. *Female Pastoral: Women Writers Re-Visioning the American South.* Knoxville: U of Tennessee P, 1991. Print.

Hatch, Peter J. *"A Rich Spot of Earth": Thomas Jefferson's Revolutionary Garden at Monticello.* New Haven: Yale UP, 2012. Print.

Hayasaki, Erika. "Seeds of Dissension Linger." *Los Angeles Times.* The Tribune Co. 31 October 2005. Web. 21 March 2006.

Hays, Samuel. *Beauty, Health, and Permanence: Environmental Politics in the United States, 1955–1985.* New York: Cambridge UP, 1987. Print.

Heffernan, William. "Concentration of Ownership and Control in Agriculture." *Hungry for Profit: The Agribusiness Threat to Farmers, Food, and the Environment.* Ed. Fred Magdoff, John Bellamy Foster, and Frederick H. Buttel. New York: Monthly Review, 2000. 61–75. Print.

Heise, Ursula. *Sense of Place and Sense of Planet: The Environmental Imagination of the Global.* New York: Oxford UP, 2008. Print.

Helphand, Kenneth. *Defiant Gardens: Making Gardens in Wartime.* San Antonio: Trinity UP, 2006. Print.

Higgs, Eric. *Nature by Design: People, Natural Processes, and Ecological Restoration.* Cambridge: MIT P, 2003. Print.

Hou, Jeffrey. *Insurgent Public Space: Guerrilla Urbanism and the Remaking of Contemporary Cities.* New York: Routledge, 2010. Print.

Hou, Shen. *The City Natural: Garden and Forest Magazine and the Rise of American Environmentalism.* Pittsburgh: U of Pittsburgh P, 2013. Print.

Howard, Ebenezer. *Garden Cities of To-morrow.* London: Sonnenschein, 1902. Print.

Hoyle, Martin. *Bread and Roses: Gardening Books from 1560 to 1960.* London: Pluto, 1995. Print.

———. "The Garden and the Division of Labour." *Vista: The Culture and Politics of Gardens*. Ed. Tim Richardson and Noel Kingsbury. London: Frances Lincoln, 2005. 21–38. Print.

Hynes, H. Patricia. *A Patch of Eden: America's Inner-City Gardeners*. White River Junction, CT: Chelsea Green, 1996. Print.

Ingram, David S. *Green Screen: Environmentalism and Hollywood Cinema*. Exeter: U of Exeter P, 2010. Print.

Jackson, Kenneth. *Crabgrass Frontier: The Suburbanization of the United States*. New York: Oxford UP, 1985. Print.

Jacobs, Jane. *The Death and Life of Great American Cities*. 1961. New York: Random, 1993. Print.

Jacoby, Karl. *Crimes against Nature: Squatters, Poachers, Thieves, and the Hidden History of American Conservation*. Berkeley: U of California P, 2005. Print.

Jefferson, Thomas. "Letter to Bernard Mcmahon." 25 April 1806. MS. *The Thomas Jefferson Papers*. Lib. of Cong. Web. 12 May 2014.

Jensen, Jens. *Siftings: The Major Portion of the Clearing and Collected Writings*. Ed. Ralph Fletcher Seymour. Chicago: Ralph Fletcher Seymour, 1956. Print.

Jonnnes, Jill. *South Bronx Rising: The Rise, Fall, and Resurrection of an American City*. New York: Fordham UP, 2002. Print.

Jordan, William, and George Lubick. *Making Nature Whole: A History of Ecological Restoration*. Washington, DC: Island, 2011. Print.

Kains, M. G. *Five Acres and Independence: A Practical Guide to the Selection and Management of the Small Farm*. 1935. 17th printing. New York: Greenberg, 1943. Print.

Kelsey, Elin, ed. "Beyond Doom and Gloom: An Exploration through Letters." *RCC Perspectives* no. 6 (2014): 5–69.

Kincaid, Jamaica. *My Garden (Book):*. New York: Farrar, 2001. Print.

Kleiman, Jordan. "Greening Fort Apache." Annual Meeting of the American Society for Environmental History. Baton Rouge, LA, 28 Feb. 2007. Conference paper.

Klindienst, Patricia. *The Earth Knows My Name: Food, Culture, and Sustainability in the Gardens of Ethnic Americans*. Boston: Beacon, 2006. Print.

Kolodny, Annette. *The Lay of the Land: Metaphor as Experience and History in American Life and Letters*. Chapel Hill: U of North Carolina P, 1975. Print.

Kosinski, Jerzy. *Being There*. New York: Harcourt, 1971. Print.

Krech, Shepard, III. *The Ecological Indian: Myth and History*. New York: Norton, 1999. Print.

Kunstler, James Howard. *The Geography of Nowhere: The Rise and Decline of America's Man-Made Landscape*. New York: Simon, 1993. Print.

Lachmund, Jens. *Greening Berlin: The Co-Production of Science, Politics, and Urban Nature*. Cambridge: MIT P, 2013. Print.

Lacy, Allen, ed. *The American Gardener: A Sampler*. New York: Farrar, 1990. Print.

Lawrence, Elizabeth. *Gardening for Love: The Market Bulletins*. Durham: Duke UP, 1987. Print.

———. *A Southern Garden*. 1942. Chapel Hill: U of North Carolina P, 1967. Print.

Lawson, Laura J. *City Bountiful: A Century of Community Gardening in America.* Berkeley: U of California P, 2005. Print.

Leighton, Ann. *American Gardens in the Eighteenth Century: "For Use or for Delight."* Boston: Houghton, 1976. Print.

Lekan, Thomas. "*Serengeti Shall Not Die*: Bernhard Grzimek, Wildlife Film, and the Making of a Tourist Landscape in East Africa." *German History* 29.2 (2011): 224–64. Print.

Lemenager, Stephanie. *American Gardens of the Nineteenth Century: "For Comfort and Affluence."* Amherst: U of Massachusetts P, 1987. Print.

———. *Early American Gardens: For Meate or Medicine.* Boston: Houghton, 1970. Print.

———. "Petro-Melancholia: The BP Blowout and the Arts of Grief." *Qui Parle* 19.2 (2011): 25–55. Print.

Leopold, Aldo. *A Sand County Almanac and Sketches from Here and There.* London: Oxford UP, 1949. Print.

Light, Andrew. "Elegy for a Garden." *Terrain.org: A Journal of the Built and Natural Environments* 15 (Fall/Winter 2004): 15 Oct. 2007. Web.

"Lights, Camera, Eviction." Editorial. *Los Angeles Times.* 14 June 2006. Web.

"Lilac Showed Very Early Activity in 2012." *USA National Phenology Network.* N.d. Web. 2 June 2014.

Lipsitz, George. *The Possessive Investment in Whiteness: How White People Profit from Identity Politics.* Philadelphia: Temple UP, 2000.

Lodge, David. *Changing Places: A Tale of Two Campuses.* 1974. London: Penguin, 2008. Print.

Long, Christian. "Mapping Suburban Fiction." *Journal of Language, Literature, and Culture* 60.3 (2013): 193–213. Print.

Los Angeles County Office of the Assessor. "4051 S Alameda St." *Property Assessment Information System.* N.d. Web. 12 Feb. 2008.

Louv, Richard. *Last Child in the Woods: Saving Our Children from Nature-Deficit Disorder.* Rev. ed. Chapel Hill, NC: Algonquin, 2008.

Malm, Andreas, and Alf Hornborg. "The Geology of Mankind? A Critique of the Anthropocene Narrative." *Anthropocene Review* 1.1 (2014): 62–69. Web. 7 November 2014.

Marranca, Bonnie, ed. *American Garden Writing: An Anthology.* New York: Taylor, 2003. Print.

Marris, Emma. *Rambunctious Garden: Saving Nature in a Post-Wild World.* New York: Bloomsbury, 2011. Print.

Martin, Justin. *Genius of Place: The Life of Frederick Law Olmsted.* Cambridge: Da Capo, 2011. Print.

Marx, Leo. *Machine in the Garden: Technology and the Pastoral Ideal in America.* 1964. New York: Oxford UP, 2000. Print.

Mavhunga, Clapperton Chakanetsa. *Transient Workspaces: Technologies of Everyday Innovation in Zimbabwe.* Cambridge: MIT P, 2014. Print.

McClaughlin, Robert L. "Post-Postmodern Discontent: Contemporary Fiction and the Social World." *Symploke* 12.1–2 (2004): 53–68. Print.

McClintock, Nathan. "Demarcated Devaluation in the Flatlands of Oakland, California." Alkon and Agyeman 89–120.
McKinley, William. "Speech at Dinner of the Home Market Club, Boston, February 16, 1899." *Speeches and Addresses of William McKinley: From March 1, 1898 to May 30, 1900*. New York: Doubleday, 1900. 185–93. Print.
Merchant, Carolyn. *Ecological Revolutions: Nature, Gender, and Science*. 2nd ed. Chapel Hill: U of North Carolina P, 2010. Print.
Mitman, Gregg. *State of Nature: Ecology, Community, and American Social Thought, 1900–1950*. Chicago: U of Chicago P, 1992. Print.
Mitchell, Henry. "On the Defiance of Gardeners." *The Essential Earthman: Henry Mitchell on Gardening*. 1981. Bloomington: Indiana UP, 2003. 1–3. Print.
M'Mahon, Bernard. *The American Gardener's Calendar; Adapted to the Climates and Seasons of the United States*. Philadelphia: B. Graves, 1806. Print.
Morales, Alfonso. "Growing Food *and* Justice: Dismantling Racism through Sustainable Food Systems." Alkon and Agyeman 149–76.
Morris, Edmund. *Ten Acres Enough*. New York: J. Miller, 1865. Print.
Morrison, Toni. *Playing in the Dark: Whiteness and the Literary Imagination*. New York: Vintage, 1993. Print.
Muir, John. *My First Summer in the Sierra*. 1911. *John Muir's Nature Writings*. Ed. William Cronon. New York: Lib. of Amer., 1997. 147–309. Print.
———. *The Story of My Boyhood and Youth*. 1912. Madison: U of Wisconsin P, 1965. Print.
Mumford, Lewis. *The Culture of Cities*. New York: Harcourt. 1938. Print.
Nabhan, Gary. *Enduring Seeds: Native American Agriculture and Wild Plant Conservation*. San Francisco: North Point, 1989. Print.
Nearing, Helen. *Loving and Leaving the Good Life*. White River Junction, VT: Chelsea Green, 1992. Print.
Nearing, Scott. *The Making of a Radical: A Political Autobiography*. 1972. White River Junction: Chelsea Green, 2000. Print.
Nearing, Scott, and Helen Nearing. *The Good Life: Helen and Scott Nearing's Sixty Years of Self-Sufficient Living*. 1954, 1979. New York: Schocken, 1989. Print.
New York City Coalition for the Preservation of Gardens v. Giuliani. Index No. 11661/97. Sup. Ct. of New York. 15 Oct. 1997. *Lexis*. Web. 7 May 2008.
New York City Environmental Justice Alliance v. Giuliani. Docket No. 99-7713. U.S. Ct. of Appeals 2nd Circuit. 13 Aug. 1999. *Lexis*. Web. 7 May 2008.
Nixon, Rob. *Slow Violence and the Environmentalism of the Poor*. Cambridge: Harvard UP, 2011. Print.
Norgaard, Kari M. *Living in Denial: Climate Change, Emotions, and Everyday Life*. Cambridge: MIT P, 2011. Print.
Norton, Brian G. "Environmental Ethics and Weak Anthropocentrism." *Environmental Ethics: An Anthology*. Ed. Andrew Light and Holmes Rolston III. Malden, MA: Blackwell, 2003. 163–73. Print.
Norwood, Vera. *Made from This Earth: American Women and Nature*. Chapel Hill: U of North Carolina P, 1993. Print.

Nussbaum, Martha C. *Poetic Justice: The Literary Imagination and Public Life.* Boston: Beacon, 1995. Print.
OECD. *Divided We Stand: Why Inequality Keeps Rising.* OECD Publishing. 2011. Web. 4 Aug. 2014.
Olmsted, Frederick Law. "Public Parks and the Enlargement of Towns." Cambridge: Riverside-Amer. Soc. Science Assoc., 1870. Print.
"Only Man's Presence Can Save Nature." *Harper's Magazine* 280.1679 (1990): 37–46. Print.
Perényi, Eleanor. *Green Thoughts: A Writer in the Garden.* New York: Random, 1981. Print.
Phillips, Dana. *The Truth of Ecology: Nature, Culture, and Literature in America.* Oxford: Oxford UP, 2003. Print.
Phillips, Sarah T. *This Land, This Nation: Conservation, Rural America, and the New Deal.* Cambridge; New York: Cambridge UP, 2007. Print.
Pollan, Michael. *Second Nature: A Gardener's Education.* New York: Dell, 1991. Print.
"Pückler and America." Ed. Sonja Duempelmann. *GHI Bulletin Supplement 4.* Washington, DC: German Historical Institute, 2007. Print.
Pückler-Muskau, Hermann L. H. *Andeutungen über Landschaftsgärtnerei.* Stuttgart: Hallberger, 1834. Print.
Pulido, Laura. "Community, Place, and Identity." *Thresholds in Feminist Geography: Difference, Methodology, Representation.* Ed. John Paul Jones, Heidi J. Nast, Susan M. Roberts. New York: Rowman, 1997. 11–28. Print.
Putnam, Jean-Marie Consigny, and Lloyd C. Cosper. *Gardens for Victory.* New York: Harcourt, 1942. Print.
Randolph, Anne Cary. Entry for August 1805. *Thomas Jefferson's Household Accounts.* Ed. Gerard W. Gawalt. Manuscript Division, Lib. of Cong. Web. 12 May 2014.
Randolph, John. *A Treatise on Gardening.* 1793. Introd. Marjorie F. Warner. Richmond: Appeals, 1924. Print.
Revkin, Andrew. "Buried Seed Vault Opens in Arctic." Dot Earth. *New York Times.* 26 Feb. 2008. Web. 18 July 2014.
Richardson, Jill. "Organic White House Garden Puts Some Conventional Panties in a Twist." *La Vida Locavore.* N.p. 28 March 2009. Web. 10 August 2009.
Riis, Jacob. *How the Other Half Lives.* 1890. New York: Dover, 2012. Print.
Robbins, Paul. *Lawn People: How Grasses, Weeds, and Chemicals Make Us Who We Are.* Philadelphia: Temple UP, 2007. Print.
Robbins, Paul, and Trevor Birkenholtz. "Turfgrass Revolution: Measuring the Expansion of the American Lawn." *Land Use Policy* 20.2 (2003): 181–94. Print.
Robinson, William. *The English Flower Garden.* 8th ed. London: John Murray, 1903. Print.
Roosevelt, Franklin Delano. "The Food Problem." Message to Congress, Nov. 1, 1943. *Vital Speeches of the Day* 10.3 (1943): 66–75. Print.
Rorty, Richard. *Contingency, Irony, and Solidarity.* Cambridge: Cambridge UP, 1989. Print.

Rosenthal, Elisabeth. "Near Arctic, Seed Vault Is a Fort Knox of Food." *New York Times.* 29 Feb. 2008. Web. 18 July 2014.

Rosenzweig, Roy, and Elizabeth Blackmar. *The Park and the People: A History of Central Park.* Ithaca, NY: Cornell UP, 1992. Print.

Schlosberg, David. *Defining Environmental Justice: Theories, Movements, and Nature.* Oxford: Oxford UP, 2007. Print.

Seed, Patricia. *Ceremonies of Possession in Europe's Conquest of the New World, 1492–1640.* New York: Cambridge UP, 1995. Print.

Shabecoff, Philip. *A Fierce Green Fire: The American Environmental Movement.* Rev. ed. Washington, DC: Island, 2003. Print.

Shi, David E. *The Simple Life: Simple Living and High Thinking in American Culture.* New York: Oxford UP, 1986. Print.

Skinner, Jonathan. "Gardens of Resistance: Gilles Clément, New Poetics, and Future Landscapes." *Qui Parle* 19 (2011): 259–74. Print.

"South Central Farmers: What we are about!" *South Central Farmers.* N.d. Web. 12 Feb. 2008.

Spirn, Anne Whiston. *The Language of Landscape.* New Haven: Yale UP, 1998. Print.

State v. City of New York. No. 2000-02038. Supreme Ct. of New York. 18 Sep. 2000.

Steinberg, Ted. *Down to Earth: Nature's Role in American History.* New York: Oxford UP, 2002. Print.

Stewart, Mart A. "Rice, Water, and Power: Landscapes of Domination and Resistance in the Lowcountry, 1790–1880." *American Environmental History.* Ed. Louis Warren. Malden, MA: Blackwell, 2003. 125–37. Print.

———. *"What Nature Suffers to Groe": Life, Labor, and Landscape on the Georgia Coast, 1680–1920.* Athens: U of Georgia P, 1996. Print.

Sweet, Timothy. *American Georgics: Economy and Environment in Early American Literature.* Philadelphia: U of Pennsylvania P, 2001. Print.

"3'-By-4' Plot of Green Space Rejuvenates Neighborhood." *The Onion.* 11 Feb. 2008. Print.

Tishler, William H. Introduction. *Writings Inspired by Nature.* By Jens Jensen. Madison: Wisconsin Historical Society P, 2012. Print.

Torres, Jose. "The Fight to Save Esperanza Garden." *All Things Considered.* National Public Radio. 24 Feb. 2000. Web. 12 May 2008.

Two Gardeners: Katharine S. White and Elizabeth Lawrence, a Friendship in Letters. Ed. Emily Herring Wilson. Boston: Beacon, 2002. Print.

Unger, Nancy C. *Beyond Nature's Housekeepers: American Women in Environmental History.* Oxford: Oxford UP, 2012. Print.

United States. Dept. of Ag. Ag. Research Service. *Plant Hardiness Zone Map.* 2012. Web. 14 Aug. 2014.

United States. Dept. of Health and Human Services. FDA. Pesticide Monitoring Program. *Fiscal Year 2012 Pesticide Report.* 2 Apr. 2015. Web. 14 June 2015.

"Urban Garden Made from Compost." *Mother Earth News.* May/June 1978. *Mother Earth News.* Web. 5 June 2014.

Urban Roots. Dir. Leila Conners and Mark MacInnis. Tree Media, 2010. Film.
USA National Phenology Network. *Nature's Notebook*. N.d. Web. 8 July 2014.
Von Hassell, Malve. *The Struggle for Eden: Community Gardens in New York City*. Westport, CT: Bergin, 2002. Print.
Walker, Alice. "In Search of Our Mothers' Gardens." 1974. *In Search of Our Mothers' Gardens*. Orlando: Harvest-Harcourt, 1983. Print.
———. *Meridian*. Orlando: Harvest-Harcourt, 1976. Print.
Warnes, Andrew. *Hunger Overcome? Food and Resistance in Twentieth-Century African American Literature*. Athens: U of Georgia P, 2004. Print.
Welty, Eudora. "The Wanderers." *The Golden Apples*. 1949. *Stories, Essays, & Memoir*. New York: Lib. of America, 1998. 515–56. Print.
Westamacott, Richard. *African-American Gardens and Yards in the Rural South*. Knoxville: U of Tennessee P, 1992. Print.
The West Philadelphia Landscape Project. MIT. N.d. Web. 1 July 2015.
White, Evelyn. "Black Women and Wilderness." *The Stories That Shape Us: Contemporary Women Write about the West: An Anthology*. Ed. Teresa Jordan and James Hepworth. New York: Norton, 1995. Print.
White, Katharine. *Onward and Upward in the Garden*. Ed. with preface by E. B. White. 1979. Boston: Beacon, 2002. Print.
Will, Brad. "Cultivating Hope: The Community Gardens of New York City." *We Are Everywhere: The Irresistible Rise of Global Anticapitalism*. New York: Verso, 2003. 134–39. Print.
Williams, Raymond. "Ideas of Nature." *Problems in Materialism and Culture: Selected Essays*. London: Verso, 1980. 67–85. Print.
Wilson, Emily Herring. *No One Gardens Alone: A Life of Elizabeth Lawrence*. Boston: Beacon, 2004. Print.
Wilson, Ernest H. *Aristocrats of the Garden*. New York: Doubleday, 1917. Print.
Wilson, Gilbert L. [and Maxi'diwiac Waheenee]. *Agriculture of the Hidatsa Indians: An Indian Interpretation*. Minneapolis: U of Minnesota, 1917. Print.
Witt, Doris. *Black Hunger: Soul Food and America*. St. Paul: U of Minnesota P, 2004. Print.
Woefle-Erskine, Cleo. *Urban Wilds: Gardeners' Stories of the Struggle for Land and Justice*. 2nd ed. Oakland: AZ, 2003. Print.
Worster, Donald. *Dust Bowl: The Southern Plains in the 1930s*. New York: Oxford UP, 1979. Print.
Wright, Richardson L. "Food of Peace [victory gardens]." *House & Garden* 87 (May 1945): 53. Print.
———. *The Gardener's Bed-Book*. 1929. New York: Mod. Lib., 2003. Print.
———. "Second Call to Victory Gardeners." *House & Garden* 83 (Jan. 1943): 31. Print.
Wulf, Andrea. *Founding Gardeners: The Revolutionary Generation, Nature, and the Shaping of the American Nation*. New York: Knopf, 2011. Print.
Yaeger, Patricia. *Dirt and Desire: Reconstructing Southern Women's Writing*. Chicago: U of Chicago P, 2000. Print.

INDEX

ACGA (American Community Gardening Association), 99, 134, 139, 157, 206n2
Adams, John, 14
Adamson, Joni, 68, 143, 169
advertisements, 40–41
aesthetics, 4, 18, 25–26, 39–50, 60–65, 76, 81–83, 99, 122–30, 154–55, 172–75, 185–86
African Americans: environmental justice movement and, 134–35, 168–70, 175–84; gardening practices of, 104–7, 116–17, 130–31, 170–75; Jim Crow South and, 115–18; land use policies and, 118–19, 204n1; memory and, 124–30; slavery and, 15, 103, 105–6, 109–11. *See also* colonialism; race; South, the
Agyeman, Julian, 69–70
Aleshire, Trish, 203n1
alienation, 200n10
Alkon, Alison, 69–70
Allen, Barbara, 130
Allen, Henry, 7, 58, 60

Allen, Will, 182–84, 188, 194
almanacs, 13
American Garden (Jensen), 28
American Garden Club, 21, 25
An American Gardener (Cobbett), 51
The American Gardener's Calendar (M'Mahon), 13–14
American Georgics (Sweet), 200n10
Anderson, Amanda, 84–87
Anthropocene (Stoermer), 196–97
anthropocentrism, 86, 90–97, 192–97
Appetite for Change (Belasco), 76
archives metaphor, 10, 15, 70, 97–102, 106, 173
Arikara, 11
Aristocrats of the Garden (Wilson), 21–22
Arnold Arboretum, 20–21
At Home in Nature (Gould), 202n6
Atlantic, 58
"At the Edge of the Woods" (Goodman), 29
Aust, Franz, 30
Austin, Mary, 114

INDEX

autobiography, 83–97

back-to-the-land movement, 49, 178.
  See also Nearing, Helen and Scott
Baez, Joan, 164
Bailey, Liberty Hyde, 12–13, 20–24,
  33–34, 47, 49, 63, 75–76, 108
Balmori, Diana, 152–55, 187
Bartram, Benjamin, 16
Bay Area Seed Interchange Library, 191
beautility, 41
Beecher, Catharine, 25
Being There (Kosinski), 8, 71, 73–84,
  188
Belasco, Warren, 76, 167
Benjamin, Walter, 123
Bennett, Michael, 110, 125, 169
Berkes, Fikret, 128
Berry, Wendell, 23, 126, 200n11, 201n17
Beverly, Robert, 104–5
Białowieża forest, 189–90
Biltmore Estate, 75, 82
Black Elk Speaks (Neihardt), 128, 204n8
Blackmar, Elizabeth, 19, 75
"Black Women and the Wilderness"
  (White), 125
Blanchan, Neltje, 44
Bloomberg, Michael, 146
*The Blossom Circle of the Year for
  Southern Gardens* (Dillon), 107–13,
  113, 116
Bodziock, Joseph, 204n4
Booth, Wayne, 86
Borsodi, Ralph, 35–38, 46–47, 50–51, 65,
  201n16
The Botany of Desire (Pollan), 91, 101
Botkin, Daniel, 1–2
Bourdieu, Pierre, 155, 185–86
boxwood hedges, 111–13
Brandt, Deborah, 137–38
bread labor, 7, 51, 55
British Empire, 173, 189
Bronx United Gardeners, 146

Brooklyn Botanical Garden, 21
bucolic futurism, 182–84
Buell, Lawrence, 4–5, 8, 187
Buffalo Bird Woman (Maxi'diwiac), 12
Bullard, Robert, 130, 139, 204n1

Cable, George Washington, 44–46
CALPIP (California Pesticide Information Portal), 202n3
Čapek, Karl, 58, 76
capitalism, 30–43, 50–56, 60, 78–79,
  92–93, 96–97
Capitalocene, 197
Carney, Judith, 103, 105
Carpenter, Novella, 10, 137, 169–70, 175,
  177–83
Carruth, Alison, 69, 170
Carson, Rachel, 24, 34–35, 45, 57, 62–65,
  114, 138, 202n9
Central Park (NYC), 18–19
Century magazine, 200n5
Changes in the Land (Cronon), 16,
  199n1
Changing Places (Lodge), 147–48
Charlotte Observer, 115
Charlotte's Web (White), 60
Cheever, John, 58
chemical warfare, 23
Cherokee, 11
Christy, Liz, 138, 179
cities: community gardens and, 3–4,
  10, 22, 105, 140–47, 169–70; dystopian portraits of, 5–6; flights from,
  50–56, 143; food system and, 33–38;
  industrial era and, 33–38, 47–50;
  land use issues in, 10, 105, 132–47,
  157–63, 185–86, 189, 204n1; organic
  cities idea and, 170, 175–84, 186,
  205n5; parks in, 18–19, 28; postindustrial conditions of, 5, 7, 131–47,
  161–63, 192–94; victory gardens
  and, 38–43
citizen science, 195–97

City Beautiful movement, 19–20, 108, 132–34
City Bountiful (Dawson), 133–34
Civilian Conservation Corps, 31–32
civil rights movement, 103, 126–27, 130–31
class: aesthetic concerns and, 4, 18, 39–50, 59–60; cultivation ideology and, 73–83, 97–101; environmentalism's issues with, 2, 20–21, 69–70, 73–83, 90–97, 139–40, 143, 147–57, 168–70, 187–88; gardening labor and, 4, 7, 9, 25–26, 33–34, 60–61, 116–17, 200n5, 203n3; gentrification and, 146–52, 162, 181; land use issues and, 10, 18–19, 22, 33, 38–39, 132–47, 157–63, 185, 189, 201n1; literature and, 46–50, 56–66; morality discourses and, 2–3, 20; nature's moral impact and, 17–18, 22–23; Pollan's work and, 69–73, 97–101; Progressive Era politics and, 19–20; race and, 132–40; urban spaces and, 5–6
Clayton, Virginia Tuttle, 21
The Clearing, 32
Clément, Gilles, 138, 190, 206n3
climate change, 130–31, 169, 188, 195
Clinton, Bill, 71–72
Cobbett, William, 51
Cohen, Liz, 48
Cold War, 50–56, 58–59, 69, 73, 75, 78–79, 88–90
Cole, Karen, 114
colonialism, 12–14, 16, 25–26, 64–65, 67–68, 101, 170–75, 189, 199n1, 205n1. *See also* British Empire; Kincaid, Jamaica; postcolonial analyses; slavery
comedy, 58, 70–71, 76, 90–91
Communitas (Goodman and Goodman), 200n15
Community Garden Mapping Project, 144

community gardens, 4, 7, 9–10, 22, 99, 104, 132–47, 152–63, 194
Community Supported Agriculture, 166–67
Concerned Citizens of South Central Los Angeles, 135, 163
Conners, Leila, 134
consumerism, 44–50, 91
Contingency, Irony, and Solidarity (Rorty), 83–86, 88
Continuing the Good Life (Nearings), 56
Cooked (Pollan), 68
Cooper, David, 89, 91–92
The Corrections (Franzen), 10, 147–52
Cosmos (Humboldt), 25
Council on the Environment of New York City, 144
Country Life, 21
Cowles, Henry Chandler, 31
Cozart, Bernadette, 6, 156
Crane, Stephen, 137
Crèvecoeur, Hector St. John de, 14
Crimes against Nature (Jacoby), 199n3
Crist, Eileen, 197
critical cultivation, 90–97, 197
Cronon, William, 15–16, 67–68, 199n1, 199n3
Crosby, Alfred, 199n1
Crutzen, Paul, 196–97
Cultivating Food Justice (Alkon and Agyeman), 70
Cultivating Victory (Gowdy-Wygant), 39, 77, 201n19
cultivation metaphor: agribusiness and, 33–38, 42–45, 52; antimodernity and, 50–56; community gardens and, 10, 132–47, 152–57; liberal irony and, 70–73; morality and, 2–3, 11–13, 15–16, 50–56, 91–97, 108–13; as political ideology, 73–83, 91–97; private property and, 33–34, 44–50; racial discourses and, 113–23; regional articulations of, 103;

cultivation metaphor (*continued*)
  social justice and, 170–75, 186; spiritual connotations of, 124–30. *See also* class; gardening and gardening metaphors; gender; justice; race
The Cultural Front (Denning), 10, 137
Culture of Cities (Mumford), 200n15
Curtis, John, 30, 32
Curtis Prairie, 3

Darwin, Charles, 23
Dawson, Laura, 133–34, 162
DDT, 7
Debs, Eugene, 202n7
deep ecology, 90
Deepwater Horizon, 130
Deloughrey, Elizabeth, 173
democracy: cultivation metaphor and, 11–25
Denning, Michael, 10, 137, 147, 162, 169–70
Desai, Karen, 174–75
Dial, 58
Didur, Jill, 174–75
Dillon, Julia Lester, 9, 103, 107–13, 116, 119
Dimock, Wai-Chee, 142, 158, 168–69
Divided We Stand (report), 206n6
Dorney, Robert, 192
Dos Passos, John, 137
Douglass, Frederick, 9, 107, 109, 125, 204nn4–5
Downing, Andrew Jackson, 16–18, 20, 25, 27, 31
Down to Earth (Steinberg), 175–76
down-to-earth appearance, 73–83
Dreiser, Theodore, 56
Drumlin Community Gardens, 166
Durkheim, Emile, 205n5
dust bowl, 31

Earle, Alice Morse, 25, 44, 112
"Earth Art" (Ukeles), 179
Earth First!, 1, 160
The Earth Knows My Name (Klindienst), 46
"Earthman" (Mitchell), 71, 76
ecocosmopolitanism, 168–75
ecocriticism, 1, 4–5, 104, 174. *See also* literature
Ecological Imperialism (Crosby), 199n1
ecological restoration, 25–33, 192–93
ecopoetics, 138
Eine andere Welt ist pflanzbar! (von der Haide), 134
EJA (Environmental Justice Alliance), 141–42, 145
"Elegy for a Garden" (Light), 159
Eliot, Charles, 19–20
El Jardín de la Esperanza, 157–63, 166, 187
Ely, Helena Rutherford, 112
emotions, 91–97, 203n2
Empson, William, 5–6, 162
enclosure, 111–13
The English Flower Garden (Robinson), 27–28
environment and environmentalism: African Americans and, 124–30; class issues and, 69, 139–40, 143, 147–52, 187–88; community gardens and, 140–67, 175–82; Ecological Indian figure and, 29–30; environmental literacy and, 136–38, 161; food system and, 33–38, 42–56, 62–63, 68–73, 146–47, 166–70, 175–82; gardening metaphor and, 1–2, 66–73, 83–90, 140; gender and, 2, 139–40, 156–57, 201n19; literary representations of, 5, 13, 73–83, 113–30, 147–52; media representations of, 73–75, 80; moral discourses and, 18–19, 23–33; slow violence and, 3, 23–24, 130–31, 188, 195; social justice and, 2–4, 8–10, 20–21, 23–24,

30, 69–83, 85–90, 130–47, 157–63, 168–86, 205n3; wilderness and, 3, 18, 22, 67–73, 90, 94, 99, 124–25, 189, 199n3. *See also* cities; gardening and gardening metaphors; justice; morality; pesticides; region
Epistle to Lord Burlington (Pope), 83
The Essential Earthman (Mitchell), 76
Ethics of Fiction (Booth), 86
Evans, Mei Mei, 169
Executive Order 12898, 71–72

Farm City (Carpenter), 10, 137, 169–70, 175, 177–83
Fassett, Norman, 31
Faulkner, William, 108
FDA (U.S. Food and Drug Administration), 45, 202
feminism, 104. *See also* gender
A Fierce Green Fire (Shabecoff), 99
First World War, 23, 40, 134, 140
Fitch, Robert, 141, 152, 161
Flight from the City (Borsodi), 35–38
FONL (Friends of Our Native Landscape), 28–32
Food Rules (Pollan), 68
food system, 33–38, 42–56, 62–63, 68–73, 146–47, 166–70, 175–82. *See also* gardening and gardening metaphors
Foreman, Dave, 1
Forest Farm, 50–56
formalism, 56–57
Founding Gardeners (Wulf), 14
fragrance, 25, 44
Franzen, Jonathan, 10, 147–52
Frémont, John C., 16
From Yard to Garden (Grampp), 43
frontier thesis, 36

Gaia hypothesis, 1
The Garden (Kennedy), 134–35, 165
Garden and Forest, 19

Garden Cities of To-Morrow (Howard), 37
Garden Club of Virginia, 15
The Gardener's Bed-Book (Sackville-West), 58
The Gardener's Year (Čapek), 58, 76
gardening and gardening metaphors: aesthetics of, 4, 33–34, 124–30; agribusiness and, 33–38; antimodernity and, 5–6, 35, 50–56, 97–101; class issues of, 3–4, 9, 20–22, 25–26, 34–35, 46–50, 56–66, 73–83, 97–101, 200n5, 203n3; colonialism and, 12–14, 16, 25–26, 64–65, 67–68, 101, 170–75, 189, 199n1, 205n1; community gardens and, 3–4, 7, 9–10, 104, 132–63, 194; environmentalism and, 1–2, 8–9, 25–33, 44–50, 90–97, 124–25, 130–31, 152–63, 182–85; gender and, 3–4, 6–7, 42–43, 77–78, 107–8, 114, 124–30, 156–57, 201n19, 206n1; intercultural exchange and, 101, 103–4, 111–13, 128–30, 204n7; intergenerational exchange and, 6–7, 69–73, 81–83, 93–98, 124–30; literary representations of, 13–17, 39–41, 56–66, 73–83, 147–52; morality and, 11–13, 25–26, 28–29, 41–43, 49–50, 70–73, 83–97, 102–7; Native American practices and, 11–12; private property ideology and, 11–12, 44–50, 72, 84–89, 110–13; racial divides of, 4, 9, 116–18, 190–91, 203n1; regionalism and, 4–7, 9, 114–23; scientific knowledge and, 13, 20–24, 34–35; sensuality and, 25, 91–97; spiritual work of, 124–30; subsistence and, 5, 7, 22, 33–43, 47–48, 50–56, 126–27, 136, 146–47; suburbia and, 7, 44–50, 97–101; vernacular gardens and, 103, 105, 119, 128. *See also* cultivation metaphor; food system; *specific authors and works*

Gardening for Love (Lawrence), 106, 114–15, 119–20, 122–24
Gardens for Victory, 41
Gardens in Winter (Lawrence), 115
Garden Week, 25
garden writers (definition), 48–49
Gates Foundation, 191
gender: aesthetics and, 4, 60–62; domestic ideologies and, 37–38; environmental movement's issues with, 2, 139–40, 156–57, 201n19; gardening metaphor and, 3–4, 42–43, 77–78, 96–98, 104–5, 107–8, 114, 116–17, 124–30, 206n1; intergenerational exchanges and, 81–83; morality discourses and, 108–13; nature's femininity and, 97–101; property issues and, 84–85; race and, 105–6, 116–17, 124–30; regionalism, 104–7. See also gardening and gardening metaphors; justice
gentrification, 146–52, 162, 181
Georgia Farmers Market Bulletin, 121
Georgics (Virgil), 13, 51
ghetto pastoral, 137, 143, 147–52, 159, 169–70
Ghost Town Farm (blog), 177, 181
Giuliani, Rudy, 141–42, 145–46
Glave, Dianne, 105, 117, 125
Glover, Danny, 164
Go Down, Moses (Faulkner), 108
Gold, Mike, 137
The Golden Apples (Welty), 122
Good Food Revolution (Allen and Wilson), 182–84
Goodman, Kenneth Sawyer, 29–30
Goodman, Paul, 51, 200n15
Gould, Rebecca, 49, 202n6
Gowdy-Wygant, Cecilia, 39, 77, 201n19
Grampp, Christopher, 43
Grant, Greg, 105
Gray, Asa, 23
Great Depression, 31

Green Guerillas, 138–39, 157
Greening of Harlem Coalition, 156–57
Green Thoughts (Perényi), 203n3
Green Thumb program, 144–46, 153, 157–58
green-washing, 152, 161–62
Grese, Robert E., 28–29
"Growing Food *and* Justice" (Morales), 183
Growing Power, Inc., 182–84, 194
Guerilla Gardening Collective, 160
Guha, Ramachandra, 67–68, 90
Guthman, Julie, 69, 77

Hanchett, Thomas W., 118
Hansen, James, 89–90
Haraway, Donna, 87, 197
Harper's, 1–2, 87, 90, 148, 189
Harrison, Elizabeth, 6, 108
Hatch, Peter, 15
Haynesworth, H. C., 113
Hays, Samuel, 202n8
Heise, Ursula, 104, 168–70
Helphand, Kenneth, 140–41, 162
Hennig, Reinhard, 206n4
Hidatsa, 12
Higgs, Eric, 30–31, 192–93
Ho-Chunk, 12
The Holy Earth (Bailey), 22–23
"Home and Office" (White), 64
Home Beautiful, 39
Home Market Club, 42
homesteading, 35–38, 136–37
Hopi, 11
Horowitz, Ralph, 163, 165
The Horticulturist, 17–18
Hortus (Bailey), 21
Hou, Jeffrey, 134, 162
Hou, Shen, 19
House and Garden, 21, 33–34, 39, 41, 65, 76
Howard, Albert, 50
Howard, Ebenezer, 37

How the Other Half Lives (Riis), 154
HPD (Housing Preservation and Development), 145
humor, 58, 70–71, 76, 90–91
Hunger Overcome? (Warnes), 203n1
Hurricane Katrina, 130

"Ideas of Nature" (Williams), 47
In Defense of Food (Pollan), 101
individualism, 99–101. *See also* consumerism; neoliberalism; Pollan, Michael; private property; public spaces; Rorty, Richard
industrial liberalism, 33–38
"In Search of Our Mothers' Gardens" (Walker), 124
interpretation, 155
irony, 8, 84–90, 98, 102–3, 148–49, 187
"Island Ecologies and Caribbean Literature" (Deloughrey), 173
Italian Villas and Their Gardens (Wharton), 200n5
"It's Morning in America" (ad), 82

Jacobs, Jane, 137
Jacoby, Karl, 199n3
James, William, 84–85
Jefferson, Thomas, 4, 11, 14–16, 20, 22–23, 37–38, 196, 199n2
Jensen, Jens, 27–32
Jens Jensen (Grese), 28–29
Jewett, Sarah Orne, 120
Jim Crow legal regime, 7, 117, 119–21
Jordan, William, 31, 200n8
justice: class and, 2–4, 132–40, 175–82, 190–91; cultivation metaphor and, 170–75; ecological restoration and, 25–33; environmental ethics and, 2–3, 46, 66–73, 77–83, 90–97, 100–101, 140–41, 159–60, 168–70, 187, 205n3; food gardening and, 3–4, 33–38, 132–63, 175–82, 194; garden writing and, 64–66, 130–31; gendered, 2, 85–90, 156–57; literary criticism and, 4–5, 88; racial, 2, 4, 126–27, 132–40, 175–82; residues of, 142, 157–58, 161, 166. *See also* cities; class; environment and environmentalism; food system; gender; pesticides; race

Kant, Immanuel, 92
Kelsey, Elin, 203n2
Kennedy, Scott Hamilton, 134–35, 165
Kincaid, Jamaica, 169–75, 187–88
King, Louisa Yeomans, 25, 61, 108
kitchen gardens, 40, 43, 103, 126, 129
Kleiman, Jordan, 161
Klindienst, Patricia, 46
Knothe, Helen, 50. *See also* Nearing, Helen and Scott
Kolbert, Elizabeth, 57
Kolodny, Annette, 4, 203n4
Kosinsky, Jerzy, 8, 71, 73–84, 188
Kowarik, Ingo, 206n3
Krech, Shepard, III, 29–30
Kunstler, James Howard, 176–77

Lacy, Allen, 115
Ladies Home Journal, 21
"The Lady of Shalott" (Tennyson), 62
La Guardia, Fiorello, 202n7
Langley, Batty, 18
Last Child in the Woods (Louv), 193
Lawn People (Robbins), 45
lawns, 39, 42–50, 94–95, 99–100, 201n1, 202n3
Lawrence, Elizabeth, 4, 6, 9, 101, 103, 106–8, 113–24, 187
Leighton, Ann (pen name of Isadore Smith), 4, 12
Leopold, Aldo, 26–27, 30–32, 100, 192, 196
Le Rouge et le Noir (Stendhal), 112
Letters from an American Farmer (Crèvecoeur), 14

Lewis and Clark Expedition, 16
LHIC (Libaw-Horowitz Investment Company), 164
liberal ironists, 8, 71–73, 83–90, 98, 102–3, 148–49, 187. *See also* Pollan, Michael; Rorty, Richard
Light, Andrew, 92, 159–60
Lipsitz, George, 77, 202n1
literacy (environmental), 136–38, 161
literature: almanacs as, 13; categories of, 7, 12–15, 64–66; ecojustice revisionism in, 4–6, 104; ethical criticism and, 84–90; formalism in, 56–57; garden writing and, 46–50, 56–66, 73–90; modernism in, 48, 56, 108; nature's representation in, 20–21; neoliberalism in, 147–52; political positionings of, 67–83; race and, 103, 124–30; regionalism and, 4–6, 102–7, 114–30, 203n1; styles of, 54–56, 64–66, 69–73, 90–91. *See also specific authors and works*
The Little Bulb (Lawrence), 115
Living the Good Life (Nearings), 6–8, 49–56
Lloyd, Colonel Edward, 110–11, 204n5
Lob's Wood (Lawrence), 115
locavores, 35, 170
Locke, John, 84, 165–66
Lodge, David, 147
Long, Christian, 147
Longenecker, G. William, 30–32
Los Angeles Times, 165
Loudon, John C. and Jane, 16–17
Louv, Richard, 193
Lovelock, James, 1
Loving and Leaving the Good Life (Nearing), 202n6
Lubick, George, 31, 200n8

Maathi, Wangari, 157
MACA (Mid America CropLife Association), 188

MacInnis, Mark, 134
Made from This Earth (Norwood), 114
Madison, James, 14–15, 20
magnolias, 127, 203n3
Making Nature Whole (Jordan and Lubick), 200n8
The Making of a Radical (Nearing), 50
Mandan, 11
market bulletins, 118–23
Maroon gardens, 105–6
Marris, Emma, 187, 189–90, 192–94
Martin, Justin, 75
Marx, Karl, 200n10
Marx, Leo, 6, 149
masculine pastoral, 104, 107–8
Massachusetts Bay Colony, 12, 15
Maxi'diwiac, 12
McCarthy, Mary, 58
McClaughlin, Robert, 151
McClintock, Nathan, 177–78
McKibben, Bill, 90
McKinley, William, 42
memory, 25, 103–4, 108, 124–30, 153, 172–75, 178, 204n6
Menominee, 11
Merchant, Carolyn, 13, 87
Meridian (Walker), 106–7, 124–30
Milwaukee Urban Gardens, 194
Mississippi Market Bulletin, 119–20
Mitchell, Henry, 71, 76, 140–41
Mitman, Gregg, 31
M'Mahon, Bernard, 12–14, 16, 76, 199n2
modernism (literary), 48
Monticello, 4, 15
Moore, Jason, 197
Morales, Alfonso, 183
morality: aesthetics and, 15–16, 18, 25; criticism and, 86–90; ecological restoration, 25–33; emotions and, 91–97; food gardening and, 33–38; gardening as metaphor for, 70–83; gender and, 108–13; literary style and, 54–56; nature's effect on, 16–17,

19, 25–26, 28–29, 34, 49–50, 59–60, 81–83; private spaces and, 37–38; racial discourses and, 26, 108; subsistence gardening and, 50–56; wilderness and, 15–16. *See also* cultivation metaphor; environment and environmentalism; gardening and gardening metaphors; justice
More Gardens! (organization), 160
Morrison, Toni, 118
Morton, Margaret, 152–55
Mother Earth News, 161
Muir, John, 22, 25–27, 32, 124
Mumford, Lewis, 37, 97, 200n15
Murdoch, Iris, 92
My Bondage and My Freedom (Douglass), 110–11, 204n5
My First Summer in the Sierra (Muir), 26
My Garden (Book) (Kincaid), 169–75, 188

Nabhan, Gary Paul, 12
Narrative (Douglass), 107, 109–10
Nash, Roderick, 124
Nashville Agrarians, 114
National Council of State Garden Clubs, 61
National People of Color Environmental Leadership Summits, 71
National Phenology Network, 196–97
National Public Radio, 158
Native Americans, 11–12, 29–30, 68, 143, 204n8
nature: back-to-the-land movement and, 49–50; class issues and, 20; ecological restoration and, 25–33; gardening metaphor and, 1–2, 12–13, 25–33, 124–25; gendering of, 97–101, 104–5; historicization of, 69–73, 75, 87–88, 97–101; human codevelopment and, 12–13, 25–33; moral impact of, 2–3, 16–17, 19–20, 22–23,

25–26, 49–50, 59–60, 81–83. *See also* cities; cultivation metaphor; environment and environmentalism; morality
Nature by Design (Higgs), 30
Nearing, Helen and Scott, 4–8, 34, 46–56, 59, 65–75, 88, 102–3, 137, 143, 170, 202nn6–7
Neihardt, John G., 128
NEJAC (National Environmental Justice Advisory Committee), 71–72
neo-Europes, 199n1
neoliberalism, 68–69, 93, 136, 140, 147–52, 168–70, 181, 191, 206n6
neopragmatism, 85–86
New Deal, 35, 38
New England, 102; histories of, 15–16
New South, 119–23
New Yorker, 8, 46, 56–66, 90, 114
New York Times, 156
Nixon, Rob, 131
Nolen, John, 118
Nordhaus, Ted, 169
Norton, Brian, 92
Norwood, Vera, 4, 114
Nussbaum, Martha, 88
NYCCPG (New York City Coalition for the Preservation of Gardens), 145–46
NYCEJA (New York City Environmental Justice Alliance), 145–46

OASIS (Open Space Greening Praxis), 144
Obama, Michelle, 77, 188, 206n1
Occupy Wall Street, 140
Odum, Howard W., 114
O'Hara, Frank, 58
Olmsted, Frederick Law, 16–20, 25, 31, 75, 108–9, 189, 193
Omnivore's Dilemma (Pollan), 68, 91, 101
The Onion, 136

"Only Man's Presence Can Save Nature" (forum), 1–2, 87, 90, 189
"On the Defiance of Gardeners" (Mitchell), 140–41
Onward and Upward in the Garden (White), 7–8, 49, 56–66
organic cities, 170, 182–84, 186, 205n5
Organic Farming and Gardening (Rodale), 201n18
Organic Gardening, 39, 65

Parmentier, André, 16, 18
A Patch of Eden (Hynes), 156–57
People's Park, 138, 147–48
"Perchance to Dream" (Franzen), 148, 151
Perényi, Eleanor, 203n3
pesticides, 7, 23–24, 34–35, 44–45, 52–54, 62–63, 97–98, 138, 182–88, 201n1, 202n3
phenology, 196–97
Philippines, 42
Phillips, Dana, 5
Phillips, Sarah T., 35
A Philosophy of Gardens (Cooper), 89, 91–92
photography, 152–54
PHZM (Plant Hardiness Zone Map), 195
Pigford v. Glickman, 204n7
Pima, 11
Pingree, Hazen S., 134
Playing in the Dark (Morrison), 118
Poetic Justice (Nussbaum), 88
Poland, 189–90
Pollan, Michael: liberal irony and, 70–73, 78, 140, 170, 187; "Only Man's Presence Can Save Nature" and, 1–2, 87, 90, 189; *Second Nature* and, 8, 66–73, 77, 81–86, 88, 90–102, 137, 173, 177, 203n3
The Possessive Investment in Whiteness (Lipsitz), 202n1

postcolonial analyses, 169–75, 189, 199n1
Powhatan, 11
pragmatism, 67–73, 83–97, 100
private property, 11–12, 21–22, 33–50, 72, 84–89, 94–113, 136–51, 157–67, 186, 202n1
Progressive Era, 6–7, 11–13, 19–20, 29–34, 50, 55–62, 74–75, 108, 118–20, 134
propaganda, 38–40
public spaces, 18–19, 108–9, 140–47, 162–67, 175–82, 186. *See also* cities; private property
Pückler-Muskau, Hermann von, 199n4

race: class and, 132–40; community gardening and, 132–40, 147–63, 194; divisions of labor and, 9, 26; environmentalism's traditional ignoring of, 2, 67–68, 71–72, 102–3, 130–31, 158, 168–75; gender and, 105–6, 116–17, 124–30; intercultural exchange and, 101; land use issues and, 4–5, 18–19, 22, 38–39, 105, 109–10, 118–19, 132–47, 157–63, 185, 189, 202n1, 204n1; morality and, 26; racialized discourses and, 113–23; slavery and, 15; in the South, 7, 9, 102–7, 116–17, 120–21, 203n1, 203n3; whiteness and, 77, 106, 120–21, 147–52, 202n1, 203n3
Radcliffe, Carolyn, 156
"Radical American Environmentalism and Wilderness Preservation" (Guha), 68
Rambunctious Garden (Marris), 189
Randolph, Anne Cary, 4
Randolph, John, 13
Reagan, Ronald, 82
"Recent Progress in American Horticulture" (Bailey), 23
region, 4–7, 9, 14, 102–30, 203n1
relief gardens, 22

residues of justice, 142, 157–58, 161, 166
Revkin, Andrew, 191
Riverside Park (Milwaukee), 192–95
Robbins, Paul, 43, 45
Robinson, William, 18, 27–28
Rockefeller Foundation, 161–62
Rodale, J. I., 50, 201n18
Roosevelt, Franklin Delano, 40
Rooted in the Earth (Glave), 125
Rorty, Richard, 8, 72, 83–86, 88
Rosenthal, Elisabeth, 191
Rosenwald, Julius, 28
Rosenzweig, Roy, 19, 75
Rosler, Martha, 179
Rosomoff, Richard, 103, 105
Rotary Centennial Arboretum, 194
Royal Botanic Gardens, 27–28

Sackville-West, Vita, 58
St. Nicholas, 202n9
Sargent, Charles Sprague, 20
SBF (South Bronx Frontier), 161–62
Schumacher, E. F., 200n15
science, 13, 16, 20, 23, 34–35, 76
Second Nature (Pollan), 8, 66–73, 77, 81–86, 88, 90–102, 137, 173, 177, 203n3
Second World War, 35, 38–43, 45, 47, 52, 134, 201n17
Seed, Patricia, 16, 205n1
seed bombs, 139, 157–63
Sellars, Peter, 73
Seneca, 11
Sense of Place, Sense of Planet (Heise), 170–71
Shabecoff, Philip, 99, 171
sharecropping, 34, 105, 183
Shellenberger, Michael, 169
Shi, David, 49–51
Sibley squash, 91, 93–97, 100–101, 203n3
Sierra Club, 22, 26
Siftings (Jensen), 27, 30

Silent Spring (Carson), 24, 34–35, 45, 57, 62–65, 138
Sinclair, Upton, 56, 202n7
Sixth Extinction (Kolbert), 57
Skinner, Jonathan, 138
slavery, 15, 103, 105–6, 110–11
Slow Food movement, 166–67, 188
slow violence, 3, 130–31, 188, 195
Small Is Beautiful (Schumacher), 200n15
Smith, Isadore (Ann Leighton), 4, 12
Smithsonian Institute, 17
Snyder, Gary, 104
Social Gospel, 59–60, 108–9
social realism, 148
solidarity, 10, 72, 83–88, 120, 170, 180–82
Some Versions of the Pastoral (Empson), 5–6
South, the, 7, 9, 101–31, 203n1
South Central Farm, 132–35, 163–67
A Southern Garden (Lawrence), 106, 113–23
Sperry, Theodore, 31
spiritual work, 124–30
Spirn, Anne Whiston, 94, 132, 136
Spitzer, Eliot, 145–46
Steinbeck, John, 56
Steinberg, Ted, 175–76, 205n5
Stendhal, 112
Stewart, Mart, 105
Stoermer, Eugene, 196–97
The Story of My Boyhood and Youth (Muir), 26
Stowe, Harriet Beecher, 25, 107
subsistence gardening, 5, 7, 33–43, 126–27. *See also* food system; gardening and gardening metaphors
suburbs: gardening practices in, 7, 44–50, 97–101; lawns of, 39, 44–50, 94–95, 99–100, 201n1, 202n3
Sukopp, Herbert, 206n3
Survey Graphic, 64

INDEX

The Survival of the Unlike (Bailey), 23
Svalbard Global Seed Vault, 190–91, 206n4
symbolic power, 152–57, 161

taste, 59–60, 65, 185–86
tenant farmers, 34
Tennyson, Alfed, Lord, 62, 65
Thaxter, Celia, 25
Thoreau, Henry David, 83, 102, 196
Thurber, James, 58
Torres, Alicia, 157–63
Torres, Jose, 158
tourism, 105
toxicity, 7, 23–24, 34–35, 44–45, 52–54, 62–63, 138, 182–88, 201n1, 202n3
Transitory Gardens, Uprooted Lives (Balmori and Morton), 152–55
Travels (Olmsted), 109
A Treatise on Gardening (Randolph), 13
"The Trouble with Wilderness" (Cronon), 68, 199n3
Trust for Public Land, 144, 146
The Truth of Ecology (Phillips), 5
Turner, Frederick, 1
Turner, Frederick Jackson, 36
Two Gardeners (Wilson), 106

Ukeles, Mierte, 179
Uncle Tom's Cabin (Stowe), 107
unselfing, 91–92
Unsettling of America (Berry), 200n11, 201n17
Updike, John, 58
urban agriculture, 169, 175–82, 186–87, 194, 205n4. *See also* cities; community gardens
Urban Ecology Center (UEC), 193–94
The Urban Farmer (business), 205n4
urban planning, 19–20, 134, 140, 143–44. *See also* cities; public spaces
Urban Roots (Conners and MacInnis), 134

UW-Madison Arbortum, 31

Vanderbilt, George Washington, 75
Vaux, Calvert, 18
vegetarianism, 51–53
Vermont, 50–56
vernacular gardens, 103, 105, 107–8, 114–15, 119, 128
Victory Garden Initiative, 194
victory gardens, 8, 38–43, 47–49, 140, 201n19
Villaraigosa, Antonio, 164–65
Virgil, 13, 51
von der Haide, Ella, 134

Walker, Alice, 4, 9, 101, 103, 106–8, 124–30
"The Wanderers" (Welty), 108, 122
"War in the Borders, Peace in the Shrubbery" (White), 58–59
Warnes, Andrew, 169, 203n1
War on Terror, 71, 88
Washington, George, 14–15, 20
Washington Post, 76
The Way We Argue Now (Anderson), 84, 86–87
Weighing In (Guthman), 77
Welty, Eudora, 108, 121–22
Westamacott, Richard, 105
West Philadelphia Landscape Project, 149
Wharton, Edith, 200n5
White, E. B., 47, 58, 202n9
White, Evelyn, 125
White, Katherine S., 4, 7–8, 46–50, 56–66, 69–75, 88, 102, 106, 114, 170, 186–87
whiteness, 77, 106, 109–10, 120–21, 147, 149–52, 202n1, 203n3. *See also* class; Lipsitz, George; race
Whitman, Walt, 19
Whole Earth Catalog, 183
"Why Bother?" (Franzen), 148, 151

Wilder, Louise Beebe, 25, 112
wilderness, 3, 15–16, 18, 22, 25–26, 67–73, 90, 94, 99, 124–25, 189, 199n3
Will, Brad, 160–62
Williams, Raymond, 47
Wilson, Charles, 182–84
Wilson, Ernest H., 21–22
Wilson, Gilbert, 12
Winfrey, Oprah, 149
Winthrop, John, 15

Women's Land Army of America, 201n19
Worster, Donald, 200n12
Wounded Knee Massacre, 204n8
Wright, Richardson, 41–42, 48, 76
Wulf, Andrea, 14–15

Yosemite, 26
Young, Arthur, 13

ROBERT S. EMMETT was born in Roanoke, Virginia, and raised in the New River Valley. He attended the University of Virginia and completed a PhD in English at the University of Wisconsin. His writing has appeared in *ISLE, Environmental Humanities,* and on the Environment & Society portal. Since 2013, he has served as the Director of Academic Programs at the Rachel Carson Center, a global hub for research in environmental humanities at the University of Munich.